SECOND EDITION

Pharmaceutical and Medical Applications of Near-Infrared Spectroscopy

PRACTICAL SPECTROSCOPY
A SERIES

1. Infrared and Raman Spectroscopy (in three parts),
   edited by Edward G. Brame, Jr. and Jeanette G. Grasselli
2. X-Ray Spectrometry, *edited by H. K. Herglotz and L. S. Birks*
3. Mass Spectrometry (in two parts), *edited by Charles Merritt, Jr. and Charles N. McEwen*
4. Infrared and Raman Spectroscopy of Polymers, *H. W. Siesler and K. Holland-Moritz*
5. NMR Spectroscopy Techniques, *edited by Cecil Dybowski and Robert L. Lichter*
6. Infrared Microspectroscopy: Theory and Applications, *edited by Robert G. Messerschmidt and Matthew A. Harthcock*
7. Flow Injection Atomic Spectroscopy, *edited by Jose Luis Burguera*
8. Mass Spectrometry of Biological Materials, *edited by Charles N. McEwen and Barbara S. Larsen*
9. Field Desorption Mass Spectrometry, *László Prókai*
10. Chromatography/Fourier Transform Infrared Spectroscopy and Its Applications, *Robert White*
11. Modern NMR Techniques and Their Application in Chemistry, *edited by Alexander I. Popov and Klaas Hallenga*
12. Luminescence Techniques in Chemical and Biochemical Analysis, *edited by Willy R. G. Baeyens, Denis De Keukeleire, and Katherine Korkidis*
13. Handbook of Near-Infrared Analysis, edited by *Donald A. Burns and Emil W. Ciurczak*
14. Handbook of X-Ray Spectrometry: Methods and Techniques, *edited by René E. Van Grieken and Andrzej A. Markowicz*
15. Internal Reflection Spectroscopy: Theory and Applications, *edited by Francis M. Mirabella, Jr.*
16. Microscopic and Spectroscopic Imaging of the Chemical State, *edited by Michael D. Morris*
17. Mathematical Analysis of Spectral Orthogonality, *John H. Kalivas and Patrick M. Lang*
18. Laser Spectroscopy: Techniques and Applications, *E. Roland Menzel*
19. Practical Guide to Infrared Microspectroscopy, *edited by Howard J. Humecki*
20. Quantitative X-ray Spectrometry: Second Edition, *Ron Jenkins, R. W. Gould, and Dale Gedcke*
21. NMR Spectroscopy Techniques: Second Edition, Revised and Expanded, *edited by Martha D. Bruch*
22. Spectrophotometric Reactions, *Irena Nemcova, Ludmila Cermakova, and Jiri Gasparic*

SECOND EDITION

Pharmaceutical and Medical Applications of Near-Infrared Spectroscopy

Emil W. Ciurczak • Benoît Igne

CRC Press
Taylor & Francis Group
Boca Raton London New York

CRC Press is an imprint of the
Taylor & Francis Group, an **informa** business

CRC Press
Taylor & Francis Group
6000 Broken Sound Parkway NW, Suite 300
Boca Raton, FL 33487-2742

First issued in paperback 2019

© 2015 by Taylor & Francis Group, LLC
CRC Press is an imprint of Taylor & Francis Group, an Informa business

No claim to original U.S. Government works

ISBN-13: 978-1-4200-8414-6 (hbk)
ISBN-13: 978-0-367-37797-7 (pbk)

Visit the Taylor & Francis Web site at
http://www.taylorandfrancis.com

and the CRC Press Web site at
http://www.crcpress.com

To my wife, Courtney, and children, for their love and support, and to all the researchers that have advanced near-infrared spectroscopy.

Benoît Igne

To my wife, Alissa, my wonderful son and daughter (Alex and Alyssa), and my special grandsons, Charlie, Kyle, and Ethan, for being my family support system.

Emil W. Ciurczak

Contents

Preface to the Second Edition

Since the completion of the first edition of this book, some major developments have shaped the field of near-infrared spectroscopy in the pharmaceutical industry. In 2002, the U.S. Food and Drug Administration (FDA) announced a new initiative to modernize its regulations of pharmaceutical manufacturing and drug quality [1]. The goals of the initiative were to incorporate risk management and quality systems concepts, and ensure that the latest scientific advances in pharmaceutical manufacturing and technology were adopted. Released in 2004, the FDA's guidance for industry titled *PAT—A Framework for Innovative Pharmaceutical Development, Manufacturing, and Quality Assurance* was intended to describe a regulatory framework that encourages the implementation and application of innovative techniques to pharmaceutical development and manufacturing [2]. Goals of process analytical technology (PAT) include enhanced understanding of the products and enhanced control of the manufacturing process. In parallel, the European Medicines Agency released notes and guidelines in 2003, 2008, 2012, and 2014 for the use of near-infrared spectroscopy for new submissions and variations [3–6]. These regulatory documents, along with the International Conference on Harmonization documents ICH Q8(R2), *Pharmaceutical Development*, ICH Q9, *Quality Risk Management*, and ICH Q10, *Pharmaceutical Quality System*, helped position near-infrared spectroscopy as a highly relevant tool for achieving control when built-in quality is preferred over quality by testing [7–9].

Consequently, a lot more interest has been given to near-infrared spectroscopy, which has been translated into myriad novel applications for the development, monitoring, and control of pharmaceutical processes. This edition takes full advantage of these developments and brings to readers an up-to-date summary of how the analytical tool, discovered by Sir William F. Herschel in 1800 and developed by Karl Norris in the 1950s and 1960s, is being applied to pharmaceutical manufacturing.

In addition, advancements in the field of functional near-infrared spectroscopy (or near-infrared spectroscopy for medicinal applications) have been summarized. Since 2001, a significant amount of work has been done in the medical field, where it is used for what it is best: noninvasive, nondestructive measurements.

References

1. M. Nasr, G. Migliaccio, B. Allen, R. Baum, and R. Branning, FDA's Pharmaceutical Quality Initiatives, *Pharm. Tech.*, 32, 54 (2008).
2. FDA, Guidance for Industry: *PAT—A Framework for Innovative Pharmaceutical Development, Manufacturing, and Quality Assurance*, 2004.
3. EMA, *Note for Guideline on the Use of Near Infrared Spectroscopy (NIRS) by the Pharmaceutical Industry and the Data Requirements for New Submissions and Variations*, 2003.
4. EMA, *Guideline on the Use of Near Infrared Spectroscopy (NIRS) by the Pharmaceutical Industry and the Data Requirements for New Submissions and Variations*, 2009.
5. EMA, *Guideline on the Use of Near Infrared Spectroscopy (NIRS) by the Pharmaceutical Industry and the Data Requirements for New Submissions and Variations*, 2012.
6. EMA, *Guideline on the Use of Near Infrared Spectroscopy (NIRS) by the Pharmaceutical Industry and the Data Requirements for New Submissions and Variations*, 2014.
7. Q8(R2): *Pharmaceutical Development*, Revision 1, ICH Harmonized Tripartite Guidelines, International Conference on Harmonization of Technical Requirements for Registration of Pharmaceuticals for Human Use, 2009.
8. Q9: *Quality Risk Management*, ICH Harmonized Tripartite Guidelines, International Conference on Harmonization of Technical Requirements for Registration of Pharmaceuticals for Human Use, 2006.
9. Q10: *Pharmaceutical Quality System*, ICH Tripartite Guidelines, International Conference on Harmonization of Technical Requirements for Registration of Pharmaceuticals for Human Use, 2007.

Preface to the First Edition

The near-infrared (NIR) region of the spectrum was discovered in 1800 by Sir William Herschel, the noted British astronomer. Considering that it took 150 years to capitalize on the region, it cannot be called an overnight success. Large bodies of work exist for the agricultural, food, textile, polymer, petroleum, and fine chemical industries. It is only in the last 15 years that serious papers have been published in the fields of pharmaceutical and medical research. During that time, the number of papers published each year has seemed to increase geometrically, leading to an explosion of references for the novice to peruse.

With this rapid growth of near-infrared spectroscopic research in the health sciences, it is time for a text such as this. The authors have combined more than 35 years of industrial and university research experience in this volume. The pharmaceutical presentation is arranged in a logical progression: theory, instrumentation, physical manipulation (blending, drying, and coating), analysis (both qualitative and quantitative), and finally, validation of the method. The varied mathematics used in NIR, called *chemometrics*, are only briefly mentioned. Detailed explanations and applications are covered in texts or chapters devoted to the subject [1–4].

In the text, we attempt to showcase the diversity of applications, not give specific "cookbook" instructions for duplication of prior work. Numerous references are given, identifying workers in the field and providing further information about their latest contributions. The applications are varied, and although the book does not include every possible citation, a representative cross section of applications and researchers in the field is presented.

All steps in the process of producing a product are assisted by NIR methods. The level of compliance with Food and Drug Administration (FDA) guidelines and the rigorousness of validation increase as the raw materials and active pharmaceutical ingredients (APIs) are mixed, granulated, and turned into a marketable product. The work in the papers cited runs the gamut from brief feasibility studies to fully validated methods ready for submission to the FDA. The amount of testing varies; however, the chemistry and principles are essentially the same for all.

The medical chapter is self-contained by design. The theory is no different from "normal" chemistry, and the instruments are often one-of-a-kind, designed by the researchers. Thus, the importance of the chapter is to give an overview of the types of applications of NIR in medicine. Major topics include: (1) diagnosis (analysis) of the chemistry of the bodily fluids, (2) chemistry of the organs, (3) imaging of major organs, and (4) chemistry of the skin. The chapter distinguishes between pediatric (pre- or postnatal) and adult applications, but only because the former deal with "wellness," while the latter apply to "illness."

Determinations thought to be in the realm of science fiction a mere decade ago are now almost routine: brain–blood oxygen, fetal maturity in utero, viability of tissue upon transplantation, and liver function. Who knows what wonders will exist in the next decade? The next year?

Emil W. Ciurczak
James K. Drennen III

References

1. H. Mark, *Principles and Practice of Spectroscopic Calibration*, John Wiley & Sons, New York, 1991.
2. H. Mark and J. Workman Jr., *Statistics in Spectroscopy*, Academic Press, New York, 1991.
3. H. Mark, Data Analysis: Multilinear Regression and Principal Component Analysis, in *Handbook of Near-Infrared Analysis*, ed. D. Burns and E. Ciurczak, Marcel Dekker, New York, 1992.
4. H.-R. Bjorsvik and H. Martens, Data Analysis: Calibration of NIR Instruments by PLS Regression, in *Handbook of Near-Infrared Analysis*, ed. D. Burns and E. Ciurczak, Marcel Dekker, New York, 1992.

chapter 1

Basic principles and theory

1.1   History

To fully appreciate the basics of near-infrared (NIR) spectroscopy, a short introduction to its origins and theory is helpful. Understanding the origin of NIR spectra may assist the practitioner with its applications.

In 1800, the English astronomer William Hershel was working on a seemingly trivial astronomical question: Which color in the visible spectrum delivered heat from the sun? He used a glass prism to produce a spectrum, while a thermometer, its bulb wrapped in black paper, was used to measure the temperature change within each color. Slight increases were noted throughout the spectrum, but they were nothing when compared with that of "pure" sunlight, i.e., all the colors combined. Going out for a meal, he left the thermometer on the table just outside the spectrum, next to the red band.

When he returned, the temperature had risen dramatically. He postulated an invisible band of light beyond (in Latin, *infra*) red, or *infrared*. Since his prism was glass and glass absorbs mid-range infrared (IR) radiation, the band was truly near infrared (short-wavelength IR). This work was duly reported [1] and essentially forgotten.

Some work was done in the near-infrared (NIR) region of the spectrum in the later portion of the 19th century, namely, by Abney and Festing [2] in 1881. They measured the NIR spectrum photographically from 700 to 1200 nm. Some workers made NIR photographs throughout the early part of this century; however, few practical uses were found for the technique. In 1881, Alexander G. Bell [3] used the NIR to heat a sample inside of an evacuated cell. Along with the sample was a sensitive microphone that he used to detect heating/expansion of the sample as it was exposed to various wavelengths of (NIR) light. This was the earliest beginnings of photoacoustic spectroscopy.

Around 1900, W. W. Coblentz used a salt prism to build a primitive infrared spectrometer [4, 5]. It consisted of a galvanometer attached to a thermocouple to detect the IR radiation at any particular wavelength. Coblentz would move the prism by a small increment, leave the room (allowing it to reequilibrate), and read the galvanometer with a small telescope. Readings would be taken for the blank and the sample. The spectrum would take an entire day to produce, and as a consequence, little work was done in the field for some time.

1.2   Early instrumentation

World War II produced the seeds for mid-range infrared instruments. Synthetic rubber produced to replace supplies lost due to the German naval presence needed to be analyzed. Industrial infrared instruments were first developed for this purpose. Postwar research showed that the mid-range region of the spectrum was more suited to structural elucidation instead of quantitative work.

With the explosion of organic synthesis, infrared spectrometers became common in almost every laboratory for the identification of pure materials and structure elucidation. With the appearance of commercial ultraviolet/visible (UV/Vis) instruments in the 1950s to complement the mid-range IRs, little was done with near infrared.

The early UV/Vis instruments, however, came with the NIR region as an add-on. Using such instrumentation, Karl Norris of the U.S. Department of Agriculture (Beltsville, Maryland) began a series of groundbreaking experiments using NIR for food applications. With a number of coauthors [6–11], he investigated such topics as blood in eggs, ripeness of melons, protein and moisture in wheat, and hardness of wheat.

It is no coincidence, then, that several companies specializing in NIR equipment should spring up in nearby communities of Maryland over the years [12]. In fairness, Dickey-John produced the first commercial NIR filter-based instrument and Technicon (now Bran + Leubbe) the first commercial scanning (grating) instrument. Available instruments and the principles of operation of each type are covered in a later chapter. However, before looking at the hardware, it is necessary to understand the theory.

1.3   Spectroscopic theory

1.3.1   Classical versus quantum models

Vibrational spectroscopy is based on the concept that atom-to-atom bonds within molecules vibrate with frequencies that may be described by the laws of physics and are therefore subject to calculation. When these molecular vibrators absorb light of a particular frequency, they are excited to a higher energy level. At room temperature, most molecules are at their rest, or zero energy, levels. That is, they are vibrating at the least energetic state allowed by quantum mechanics [13]. The lowest, or fundamental, frequencies of any two atoms connected by a chemical bond may be roughly calculated by assuming that the band energies arise from the vibration of a diatomic harmonic oscillator (Figure 1.1) and obey Hooke's law:

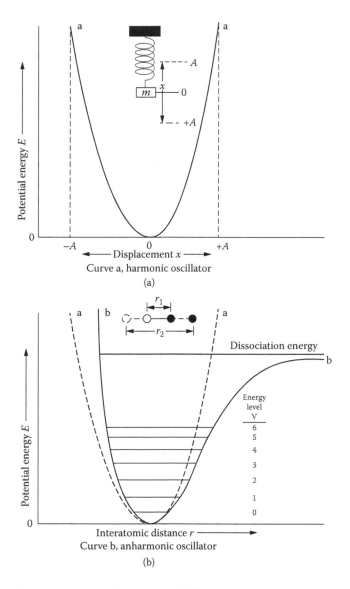

Figure 1.1 Potential energy diagrams. (a) Plot of energy versus distance for an ideal harmonic oscillator. (b) Plot of energy versus distance for a nonideal or real-life oscillator.

$$\nu = \frac{1}{2}\pi\sqrt{\frac{k}{\mu}} \qquad (1.1)$$

where ν is vibrational frequency, k is classical force constant, and μ is reduced mass of the two atoms.

This works well for the fundamental vibrational frequency of simple diatomic molecules, and is not too far from the average value of a two-atom stretch within a polyatomic molecule. However, this approximation only gives the average or center frequency of the diatomic bond. In addition, one might expect that since the reduced masses of, for example, C–H, O–H, and N–H are 0.85, 0.89, and 0.87, respectively (these constitute the major absorption bands in the near-infrared spectrum), the ideal frequencies of all these pairs would be quite similar.

In actual molecules, the electron withdrawing or donating properties of neighboring atoms and groups determine the bond strength and length, and thus the frequency. While an average wavelength value is of little use in structural determination or chemical analyses, these species-specific differences are what give rise to a substance's spectrum. The k values (bond strengths) vary greatly and create energy differences, which can be calculated and utilized for spectral interpretation.

1.3.2   Fundamental frequencies

Unlike the classical spring model for molecular vibrations, there is not a continuum of energy levels. Instead, there are discrete energy levels described by quantum theory. The time-independent Schroedinger equation

$$-\frac{h^2\partial^2\Psi(x)}{2m\partial(x)^2} + V(x)\Psi(x) = E\Psi(x) \qquad (1.2)$$

is solved using the vibrational Hamiltonian for a diatomic molecule.

Somewhat complicated values for the ground state ($\nu = 0$) and succeeding excited states are obtained upon solving the equation. A simplified version of these levels may be written for the energy levels of diatomic molecules,

$$E_\nu = \left(\nu + \frac{1}{2}\right)\frac{h}{2}\pi\sqrt{\frac{k}{\mu}} \qquad (\nu = 0, 1, 2, \ldots) \qquad (1.3)$$

where the Hooke's law terms may still be seen. Rewritten using the quantum term $h\nu$, the equation reduces to

$$E_v = \left(v + \frac{1}{2}\right)hv \qquad (v = 0, 1, 2, \ldots) \tag{1.4}$$

In the case of polyatomic molecules, the energy levels become quite numerous. To a first approximation, one can treat such a molecule as a series of diatomic, independent, harmonic oscillators. The equation for this case can be generalized as

$$E(v1, v2, v3, \ldots) = \sum_{i=1}^{3N-6}\left(v_i + \frac{1}{2}\right)hv \tag{1.5}$$

$$(v1, v2, v3, \ldots = 0, 1, 2, \ldots)$$

In any case where the transition of an energy state is from 0 to 1 in any one of the vibrational states (v1, v2, v3, ...), the transition is considered a fundamental and is allowed by selection rules. Where the transition is from the ground state to $v_i = 2, 3, \ldots$, and all others are zero, it is known as an overtone. Transitions from the ground state to a state for which $v_i = 1$ and $v_j = 1$ simultaneously are known as combination bands. Other combinations, such as $v_i = 1$, $v_j = 1$, $v_k = 1$ or $v_i = 2$, $v_j = 1$, etc., are also possible.

In the purest sense, overtones and combinations are not allowed, but they do appear as weak bands due to anharmonicity or Fermi resonance.

1.3.3   Anharmonicity and overtones

In actual practice, even the so-called ideal harmonic oscillator has limits. A graphic demonstration may be seen using the "weight from the ceiling" model. As the weight approaches the point at which the spring is attached to the ceiling, real compression forces are fighting against the bulk of the spring (often neglected in simple calculations). As the spring stretches, it eventually reaches a point where it loses its shape and fails to return to its original coil. This so-called ideal case is shown in Figure 1.1a. When the extreme limits are ignored, the barriers at either end of the cycle are approached in a smooth and orderly fashion.

Likewise, in molecules the respective electron clouds of the two bound atoms as well as the charges on the nuclei limit the approach of the nuclei during the compression step, creating an energy barrier. At the extension of the stretch, the bond eventually breaks when the vibrational energy level reaches the dissociation energy. Figure 1.1b demonstrates this effect. The barrier for decreasing distances increases at a rapid rate, while the barrier at the far end of the stretch slowly approaches zero.

As can be seen from the graphic representation, the energy levels in the anharmonic oscillator are not equal. The levels are slightly closer as the energy increases. This phenomenon can be seen in the equation

$$Ev = \left(v + \frac{1}{2}\right)h\omega_e - \left(v + \frac{1}{2}\right)^2 \omega_e\chi_e + \text{higher terms} \tag{1.6}$$

where

$$\omega_e = \left(\frac{1}{2}\right)\pi\sqrt{\frac{K_e}{\mu}}$$

is vibrational frequency, $\omega_e\chi_e$ is anharmonicity constant, K_e is harmonic force constant, and μ is reduced mass of the two atoms. (K_e is approximately 5×10^5 dynes/cm for single bonds, 10×10^5 for double bonds, and 15×10^5 for triple bonds.) For practical purposes, the anharmonicity value is usually seen to be between 1 and 5%. Thus, the first overtone of a fundamental of 3500 nm would have a range of

$$v = \frac{3500}{2} + 3500(0.01, \ 0.02, \ \ldots) \tag{1.7}$$

Depending on structural or ambient conditions, the numerical value may range from 1785 to 1925 nm for this one example. However, it would generally appear at 3500/2 plus a relatively small shift to a longer wavelength.

The majority of overtone peaks seen in a NIR spectrum arise from the X–H stretching modes (O–H, C–H, S–H, and N–H) because of energy considerations. Also, as quantum mechanically forbidden transitions, the overtones are routinely between 10 and 1000 times weaker than the fundamental bands. Thus, a band arising from bending or rotating atoms (having weaker energies than vibrational modes) would have to be in its third or fourth overtone to be seen in the near-infrared region of the spectrum.

For example, a fundamental carbonyl stretching vibration at 1750 cm^{-1} or 5714 nm would have a first overtone at approximately 3000 nm, a weaker second overtone at 2100 nm, and a third very weak overtone at 1650 nm. The fourth overtone, at about 1370 nm, would be so weak as to be analytically useless. (These values are based on a 5% anharmonicity constant.)

1.3.4 Combination bands

Another prominent feature of the near-infrared spectrum is the large number of combination bands. In addition to the ability of a band to be produced at two or three times the frequency of the fundamental, there is a tendency for two or more vibrations to combine (via addition or subtraction of the energies) to give a single band. A simple system containing combination bands is the gas SO_2 [1]. From the simple theory of allowed bands

$$\text{Number of bands} = 3N - 6 \tag{1.8}$$

there should be three absorption bands. These are the symmetric stretch (found at 1151 cm^{-1}), the asymmetric stretch (found at 1361 cm^{-1}), and the O–S–O bend (found at 519 cm^{-1}).

The three bands mentioned are the allowed bands according to group theory. However, while these are all the bands predicted by basic group theory, bands also appear at 606, 1871, 2305, and 2499 cm^{-1}. As was discussed earlier, for the molecule to acquire energy to be promoted from the ground state to a second level (first overtone) or higher is impossible for a harmonic oscillator. For the anharmonic oscillator, however, a first overtone of twice the frequency of the symmetric stretch is possible. This band occurs at 2305 cm^{-1}, with the 3 cm^{-1} difference arising from an exact doubling of the frequency accounted for by the anharmonicity.

This still leaves three bands to be explained. If the possibility exists that two bands may combine as $v_a - v_b$ or $v_a + v_b$ to create a new band, then the remaining three bands are easy to assign arithmetically. The total bands for SO_2 can be assigned as seen in Table 1.1.

In like manner, any unknown absorption band can, in theory, be deduced from first principles. When the bands in question are C–H, N–H, and O–H (4000–2500 cm^{-1}), the overtones and combinations make up much of the NIR spectrum.

Table 1.1 Band Assignments for the Infrared Spectrum of Sulfur Dioxide

v (cm^{-1})	Assignment
519	v_2
606	$v_1 - v_2$
1151	v_1
1361	v_3
1871	$v_2 + v_3$
2305	$2v_1$
2499	$v_1 + v_3$

Source: R. S. Drago, *Physical Methods in Chemistry*, W. B. Saunders, Philadelphia, 1977.

1.3.5   Fermi resonance

The vibrational secular equation deals with the fundamental vibrational modes. The general method for F and G matrices is written as

$$|FG - E\lambda| = 0 \qquad (1.9)$$

where F is the matrix of force constants, bringing the potential energies of the vibrations into the equation; G is a matrix that involves the masses and certain spatial relationships and brings kinetic energies into the equation; E is a unit matrix; and λ brings the frequency, v, into the equation and is defined as

$$\lambda = 4\pi^2 c^2 v^2 \qquad (1.10)$$

If the atomic masses in the G matrix elements are expressed in atomic mass units and the frequencies expressed in wavenumbers, then the equation may be written as

$$\lambda = 5.8890 \times 10^{-2} v^2 \qquad (1.11)$$

This may be compared with the equation obtained by treating a diatomic molecule as a harmonic oscillator,

$$f \cdot \mu^{-1} - \lambda = 0 \qquad (1.12)$$

where λ is defined as in Equation (1.11), f is the force constant between atoms, and μ is the reduced mass of the atoms.

For a polyatomic molecule, there are $3N - 6$ energy levels for which only a single vibrational quantum number is 1 when the rest are zero. These are the fundamental series, where the elevation from the ground state to one of these levels is known as a fundamental. The usual secular equation ignores overtones and combinations, neglecting their effect on any fundamental.

There are, however, occasions where an overtone or combination band interacts strongly with a fundamental. Often this happens when two excitations give states of the same symmetry. This situation is called Fermi resonance [15] and is a special example of configuration interaction. This phenomenon may occur for electronic transitions as well as vibrational modes.

We may assume the frequencies of three fundamentals, v_i, v_j, and v_k, to be related as follows:

$$v_i + v_j = v_k \qquad (1.13)$$

where the symmetry of the doubly excited state

$$\Psi_{ij} = \Psi_i(1)\Psi_j(1)\prod_{I \neq i,j}\Psi_1(0) \tag{1.14}$$

is the same as the singly excited state

$$\Psi_k = \Psi k(1)\prod_{I \neq k}\Psi_1(0) \tag{1.15}$$

This is the case where the direct product representation for the *i*th and *j*th normal modes is or contains the irreducible representation for which the *k*th normal mode forms a basis. These excited states, having the same symmetry, interact in a manner that may be described by the secular equation as

$$\begin{vmatrix} (v_i + v_j) - v & W_{ij,k} \\ W_{ij,k} & v_k - v \end{vmatrix} = 0 \tag{1.16}$$

The roots, v, are then the actual frequencies with the magnitude of the interactions given by the equation

$$W_{ij,k} = \Sigma\Psi_{ij}\,W\Psi_k\,d\tau \tag{1.17}$$

where the interactional operator, W, has a nonzero value because the vibrations are anharmonic. Since it is totally symmetric, the two states, ψ_{ij} and ψ_k, must belong to the same representation for the integral to have a finite value.

One consequence of these equations is that one of the roots of the secular equation will be greater than $(v_i + v_j)$ or v_k, while the other is less than either mode. The closer the two vibrations exist initially, the greater their eventual distance in the spectrum. Another result is a sharing of intensity between the two bands. While normally a combination band is weaker than the fundamental, in Fermi resonance the two intensities become somewhat normalized.

An example of a molecule displaying Fermi resonance is carbon dioxide. The O–C–O bending mode is seen at 667 cm^{-1} in the IR. The overtone (in Raman spectroscopy) is expected near 1330 cm^{-1}. However, the symmetric C–O stretching mode has approximately the same frequency. The result is that the observed spectrum contains two strong lines at 1388 and 1286 cm^{-1} instead of one strong and one weak line at 1340 and 1330 cm^{-1}, respectively.

While readily apparent in the better-resolved mid-range IR, this effect is somewhat difficult to observe in the NIR region, often hidden under the broad, overlapping peaks associated with this region.

Adding Fermi resonance to the overlapping combination and overtone bands already seen in the NIR, it is easy to understand why the region was dismissed for so many years as unusable. With current computers and chemometric software, however, these broad, undistinguished bands are merely nonesthetic while eminently usable. Three texts exist that are entirely devoted to the theory and application of NIR [16–18]. These may be used to supplement the information in this chapter.

1.4   Light–particle interaction

The previous sections covered the light–molecule interactions. However, there is a lot more to near-infrared spectroscopy than molecular scale phenomena. In dilute, nonscattering liquid systems, the Beer–Lambert law describes the absorption of the light:

$$A = \varepsilon l c \tag{1.18}$$

where ε is the molar absorptivity, l is the pathlength, and c the concentration of an analyte.

When measurements are performed in transmittance (the light is sent through the sample and the light not absorbed is measured), Beer's law is expressed as follows:

$$A = -log\left(\frac{I}{I_0}\right) \tag{1.19}$$

where I is the light transmitted by the sample over I_0, the light transmitted by an empty cuvette. This expression can be converted to the more familiar transmittance term (T) as follows:

$$A = \varepsilon l c = -log\left(\frac{I}{I_0}\right) = log\left(\frac{1}{T}\right) \tag{1.20}$$

When the medium is not a dilute solution, but a concentrated solution, a colloid, a semisolid, or a solid made of compacted particles, the absorption of the light deviates from Beer's law. Beer's law assumes that when a photon encounters an absorbing particle, it will be either absorbed or transmitted. However, light–particle interaction can also generate a scattering event (forward or backward scatter) when not reflected. In a particulate sample, multiple scattering events can occur, thus generating a distribution of pathlengths, larger than the original sample thickness. As the effective pathlength increases, the transmitted radiation will become more diffuse and photons scattered multiple times can come back in the direction of the light source (diffuse reflectance) or transmit through the sample (diffuse reflectance).

Diffuse reflectance is arguably the phenomenon that has made NIR so popular. It allows the interrogation of a sample without having to measure the amount of light that is transmitted, thus enabling noninvasive measurements. When employing diffuse reflectance as the measurement geometry, I_0 is replaced by the measurement of the light in a situation where no light is absorbed (reflected by a 99% Spectralon® or ceramic), and I is the light that is diffusely reflected by the sample. Absorbance is then measured as

$$A = -log\left(\frac{I}{I_0}\right) = log\left(\frac{1}{R}\right) \tag{1.21}$$

A lot of care is given in designing the detection geometry to avoid specular reflection (the reflection of light that has not interacted with the sample).

Dahm and Dahm [19] provide other deviations from Beer's law (no change in the absorbing power of any species due to interactions, using mass to prepare sample over volume fraction, and the use of nonmonochromatic light) and ask the following question: Is there a Beer's law for scattering samples? Authors describe experimental setup (very thin samples and very small detectors) for which Beer's law applied. Mark et al. published a very interesting paper on the deviation of Beer's law when mass is used to prepare samples over volume fraction [20].

A significant amount of work has come into deriving theoretical models that can be used to explain the light propagation in turbid (scattering) media. A summary of the initial work can be found in Birth and Hecht [21]. While originally considered a surface phenomenon (reflection from external surface), Seeliger described the radiation as penetrating particles where photons can be either absorbed or returned to the surface (elastic scatter). The Mie theory, which describes the scattering based on a spherical particle model, has been suggested to be the theory most relevant to NIR [22]. It is based on first principles (electromagnetic wave concept and Maxwell equations) and described in the following equation:

$$\frac{I_S}{I_0} = \frac{\lambda^2}{8\pi^2 R^2}\left(i_1 + i_2\right) \tag{1.22}$$

where for a wavelength λ, I_s is the intensity of the scattered radiation at distance R. i_1 and i_2 are the complex functions of the angle of scattered radiation.

This equation explains why as wavelength increases, the scattering intensity will decrease. This relationship will be described later in this book to explain how physical properties of samples can be modeled by NIR spectra.

As described by Griffiths and Dahm [23], a significant amount of work has been done based on a radiation transfer equation. In simple form, such an equation could be written as follows: the change in intensity of the light, I, at a particular wavelength through a sample is a function of its density ρ, pathlength l, and attenuation coefficient κ, as described by the following equation:

$$-dI = \kappa \rho I l \tag{1.23}$$

A more general form of this equation for plane-parallel layers is the following:

$$\mu \frac{dI(\tau,\mu)}{d\tau} = -I(\tau,\mu) + \frac{1}{2}\omega_0 \int_{-1}^{1} p_0(\mu,\mu')I(\tau,\mu')d\mu' \tag{1.24}$$

where μ is the cosine of the angle of the inward surface normal, μ' is the cosine of the angle of the outward surface normal, $d\tau$ is the change of optical thickness, I is the intensity of the light, ω_0 is the albedo (defined in the next paragraph), and $p_0(\mu,\mu')$ is the scattering phase function (the probability of scatter).

The albedo is defined as

$$\omega_0 = \frac{\sigma}{\sigma + \alpha} \tag{1.25}$$

where σ and α are respectively the scattering and absorption coefficients.

Thus, three optical parameters need to be estimated to solve the radiation transfer equation:

- Scattering coefficient: Defined as $\rho\sigma_s$, it is the product of the scattering cross section per unit volume.
- Absorption coefficient: Defined as $\rho\sigma_a$, it is the product of the absorption cross section per unit volume.
- Scattering phase function: Describes the fraction of light energy incident on a scattering particle that gets scattered.

Solving the radiation transfer equation is then the challenge of explaining the behavior of light in turbid media. By defining assumptions (that are beyond the scope of this chapter), the radiation transfer equation can be simplified and solved by different approaches:

- Schuster solution [24]: It consists in summarizing the radiation transfer by forward and reverse light flux with respect to the incident light. After derivation, the following equation is obtained:

$$R_\infty = \frac{1 - \sqrt{\left(\dfrac{k}{k+2s}\right)}}{1 + \sqrt{\left(\dfrac{k}{k+2s}\right)}} \qquad (1.26)$$

where k is the absorption coefficient defined as $2\alpha/(\alpha + \sigma)$, s is the scattering coefficient defined as $\sigma/(\alpha + \sigma)$, and R_∞ is the reflectance for infinite depth.

This expression can be rewritten as

$$\frac{(1 - R_\infty)^2}{2R_\infty} = \frac{k}{s} = \frac{2\alpha}{\sigma} \qquad (1.27)$$

- Kubelka–Munk solution [25]: While the Schuster solution is for one particle, Kubelka–Munk generalized the absorption and scattering phenomena to the whole sample (K and S) and solved the radiation transfer equation in a different way than Schuster. More information about the derivation is provided in Griffiths and Dahm [23]. The solution is

$$R_\infty = \frac{1}{1 + \dfrac{K}{S} + \sqrt{\left(\dfrac{K}{S}\right)^2 + \dfrac{2K}{S}}} \qquad (1.28)$$

which can be rewritten as

$$\frac{(1 - R_\infty)^2}{2R_\infty} = \frac{K}{S} \qquad (1.29)$$

Dahm and Dahm [22] have shown that both $\log(1/R)$ and the Kubelka–Munk transformation are nonlinear functions of concentration. The $\log(1/R)$ transformation is nonlinear because the absorbance coefficient is considered the additive sum of the absorbance coefficients of all absorbing species in the sample, and it does not consider that scattering varies as a function of wavelength. The nonlinearity of the Kubelka–Munk relationship is due to the erroneous assumption that the scattering coefficient, S, is assumed in practice to be constant, and independent of the level of absorption.

Dahm and Dahm recently proposed a new solution to the light propagation in turbid media based on a series of parallel infinite planes [19]. The solution takes into account the wavelength dependence between absorbance and scattering. Considering layers of planes, a new term is to be considered: remittance or the light reflected back toward the light source. They derived the following equation known as the absorbance/remission function:

$$A(R,T) = \frac{\left[\left(1-R^2\right)-T^2\right]}{R} \tag{1.30}$$

where A is the absorbance as a function of the remittance R and transmittance T. The Dahm equation is more linear than Kubelka–Munk and $\log(1/R)$ but requires the measurement of the remitted light, which requires a specific instrumental setup.

Finally, other solutions to the radiation transfer equation have been proposed using simulation to calculate the absorption and scattering coefficients. An example of simulation framework has been proposed by Wang et al. [26].

References

1. W. Herschel, Experiments on the Refrangibility of the Invisible Rays of the Sun, *Phil. Trans. Royal Soc.*, 90, 225 (1800).
2. W. Abney and E. R. Festing, On the Influence of the Atomic Grouping in the Molecules of Organic Bodies on Their Absorption in the Infra-Red Region of the Spectrum, *Phil. Trans. Royal Soc.*, 172, 887 (1881).
3. A. G. Bell, Upon the Production of Sound by Radiant Energy, *Philos. Mag.*, 11, 510 (1881).
4. W. W. Coblentz, Preliminary Communication on the Infrared, *Astrophys. J.*, 20, 1, 207–223 (1904).
5. W. W. Coblentz, *Investigation of Infrared Spectra—Part 1*, Carnegie Institution, Washington, DC, 1905.
6. I. Ben-Gera and K. H. Norris, Determination of Moisture Content in Soybeans by Direct Spectrometry, *Isr. J. Agr. Res.*, 18, 125 (1968).
7. I. Ben-Gera and K. H. Norris, Direct Spectrometric Determination of Fat and Moisture in Meat Products, *J. Food Sci.*, 33, 64 (1968).
8. K. H. Norris, R. F. Barnes, J. E. Moore, and J. S. Shenk, Predicting Forage Quality by Infrared Reflectance Spectroscopy, *J. Anim. Sci.*, 43, 889 (1976).
9. W. L. Butler and K. H. Norris, The Spectrophotometry of Dense Light-Scattering Material, *Arch. Biochem. Biophys.*, 87, 31 (1960).
10. D. R. Massie and K. H. Norris, The Spectral Reflectance and Transmittance Properties of Grain in the Visible and Near Infrared, *Trans. Amer. Soc. Agricul. Eng.*, 8, 598 (1965). 8, 598 (1965).

11. K. H. Norris and W. L. Butler, Techniques for Obtaining Absorption Spectra on Intact Biological Samples, *IRE Trans. Biomed. Elec.*, 8, 153 (1961).

12. Pacific Scientific (now Metrohm NIR Systems), Silver Spring, MD; TRABOR (later Brimrose), Baltimore; Infrared Fiber Systems, Silver Spring, MD.

13. J. D. Ingle Jr. and S. R. Crouch, *Spectrochemical Analysis*, Prentice Hall, Englewood Cliffs, NJ, 1988.

14. R. S. Drago, *Physical Methods in Chemistry*, W. B. Saunders, Philadelphia, 1977.

15. F. A. Cotton, *Chemical Applications of Group Theory*, 3rd ed., John Wiley & Sons, New York, 1990.

16. B. G. Osborne and T. Fearn, *Near-Infrared Spectroscopy Food Analysis*, John Wiley & Sons, New York, 1986.

17. P. Williams and K. H. Norris, *Near-Infrared Technology in the Agricultural and Food Industries*, American Association of Cereal Chemists, St. Paul, MN, 1987.

18. D. A. Burns and E. W. Ciurczak, *Handbook of Near-Infrared Analysis*, Marcel Dekker, New York, 1992.

19. D. J. Dahm and K. D. Dahm, *Interpreting Diffuse Reflectance and Transmittance*, NIR Publications, Chichester, 2007.

20. H. Mark, R. Rubinovitz, D. Heaps, P. Gemperline, D. J. Dahm, and K. D. Dahm, Comparison of the Use of Volume Fractions with Other Measures of Concentration for Quantitative Spectroscopic Calibration Using the Classical Least-Squares Method, *Appl. Spectrosc.*, 64, 1006 (2010).

21. G. S. Birth and H. G. Hecht, The Physics of Near-Infrared Reflectance, in *Near-Infrared Technology in the Agricultural and Food Industries*, 1st ed., ed. P. C. Williams and K. H. Norris, American Association of Cereal Chemists, St. Paul, MN, 1987, p. 1.

22. D. J. Dahm and K. D. Dahm, The Physics of Near-Infrared Scattering, in *Near-Infrared Technology in the Agricultural and Food Industries*, 2nd ed., ed. P. Williams and K. Norris, American Association of Cereal Chemists, St. Paul, MN, 2001, p. 1.

23. P. R. Griffiths and D. J. Dahm, Continuum and Discontinuum Theories of Diffuse Reflections, in *Handbook of Near-Infrared Analysis*, 3rd ed., ed. D. A. Burns and E. W. Ciurczak, Taylor and Francis, Boca Raton, FL, 2008.

24. A. Schuster, Radiation through a Foggy Atmosphere, *Astrophys. J.*, 21, 1 (1905).

25. P. Kubelka and F. Munk, Ein Beitrag zur Optik der Farbanstriche, *Z. Tech. Physik*, 12, 593 (1931).

26. L. H. Wang, S. L. Jacques, and L. Q. Zheng, MCML—Monte Carlo Modeling of Light Transport in Multi-Layered Tissues, *Comput. Methods Programs Biomed.*, 47, 131 (1995).

chapter 2

Instrumentation

Perhaps the strongest impetus for the development in near-infrared (NIR) spectroscopy in pharmaceutical and medical applications has been the explosive growth in both the types and sophistication of NIR equipment and software. Established manufacturers develop new NIR equipment yearly, and new manufacturers of NIR equipment appear regularly. This chapter deals primarily with the general types of equipment in existence, not necessarily the manufacturers themselves. Lists of manufacturers are available in various trade journals and on the Internet.

This chapter discusses the physics of each instrument type, the strengths and weaknesses of each, and a few manufacturers that produce them. It is meant as a first step toward choosing an instrument, not the last word on which one to purchase.

One underlying principle to be followed in purchasing a NIR instrument that should be adhered to is simply this: first determine your application, then purchase an appropriate instrument [1]. All too often an analyst is enamored by a particular piece of equipment and purchases it, and then looks for an application. This is a proven formula for failure. As a rule of thumb, the particular application should be clearly defined and specific objectives identified. These vital steps must be identified and the location of the test determined (in-line, at-line, or in the laboratory). At that time, instrument manufacturers should be contacted, feasibility studies conducted, references checked, and reliability of any instrument determined. Then and only then should an instrument be purchased. Most instruments available today will be suitable for a variety of applications, but choosing an inappropriate instrument for a particular application surely dooms the analyst to failure. Analysts will often erroneously assume that NIR is at fault in such situations, rather than consider that they have selected an improper instrument.

In all choices, the underlying concept should be contained in two words: necessity and sufficiency. As an example, purchasing a complex scanning instrument solely for moisture determination is a twofold mistake. It increases initial costs in hardware, training, and validation and may increase the maintenance burden due to more significant instrument complexity relative to a single-application instrument or sensor. In other words, does the instrument I am considering cover the necessary range, does it provide sufficient resolution and signal-to-noise ratio, and does the calibration effort produce sufficient payback to warrant the work? With these questions in mind, the descriptions that follow, starting with simple tools and progressing to more complex instruments, should help with your decision.

2.1   Filter-based instruments

The earliest commercial NIR instrumentation was based upon filters. (Karl Norris used a UV/Vis/NIR scanning Cary instrument, although that instrument was not designed for NIR per se.) A filter is a window that allows a particular slice of the spectrum to pass through (a bandpass filter) or blocks all wavelengths below or above a certain frequency (edge and cutoff filters).

In an interference filter, a transparent dielectric spacing material separates two partially reflective windows. This conformation forms a Fabry-Pérot filter, allowing a specific set of wavelengths to pass. The outer windows are constructed of materials with higher refractive index than the center spacer, which determines the central wavelength via its thickness.

The equation describing the central wavelength is

$$\lambda_t = \frac{2n_\sigma t_\sigma \cos\alpha}{m} \tag{2.1}$$

where λ_t is wavelength of maximum transmittance, n_σ is refractive index of the dielectric spacer, t_σ is the thickness of the dielectric spacer (in microns), α is the angle of incidence of the impinging light, and m is the order number of the interference (0, 1, 2, …). Usually, the more expensive the filter, the narrower the band.

Cutoff and bandpass filters are often designed using coatings. Where the particular wavelengths distributed in a Gaussian manner about a center wavelength are allowed to pass, a bandpass filter is the result. The coating absorbs wavelengths above and below the nominal wavelength of the filter. These are often more expensive and not easily reproducible. The coating may also lose efficiency with time. Equations based on these chemical filters often need to be regenerated when the filters are replaced.

A basic instrument designed for a single filter is shown in Figure 2.1. Modern instruments are usually equipped with moving filter wheels (inserted in light path where the simple filter is seen in the figure) and may contain 40 or more filters. These filter wheels can rotate at high speeds and produce quality data for any particular sample. In general, these instruments are not much slower than a grating-based device. The larger number of filters in the later models allows more complex models to be generated.

The basic paradigm of a filter instrument is that many applications may be run using a few select wavelengths. Experience has shown that most organic materials have common moieties, i.e., C–H, O–H, and N–H

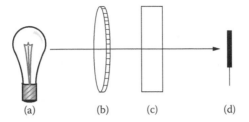

Figure 2.1 A simple filter-based spectrometer: (a) source, (b) interference filter, (c) sample, and (d) detector.

bands. These are found both as higher overtones of fundamental IR bands and in combination with other chemical species (e.g., N–H with C=O in an amide function). Thus, most early (agricultural) analyses for moisture, protein, oil, and starch used three or four filters for all these measurements, merely varying the coefficients before the individual wavelengths.

Dickey-John introduced the first commercial filter instrument in 1971 at the Illinois State Fair. Technicon first licensed the patent for its own instruments and then proceeded to produce its own design as well as a rugged scanning grating type. Pacific-Scientific (FOSS NIRSystems and now Metrohm NIRSystems) added a twist: a tilting filter instrument that generated spectra similar to that of a scanning instrument. This edition was used until commercial scanning (diffuse reflection) instruments were introduced. There are still a sizable number of manufacturers supplying filter-based instruments. In fact, many of the newer, start-up companies in the last decade offer single-analyte, filter-based equipment.

2.2 Scanning grating monochromators

The workhorse of the field for many years, (holographic) grating monochromators are still among the largest selling type of NIR instruments today (Figure 2.2). Replacing the prism instruments of the 1950s and 1960s, gratings are easily made, inexpensive, and long-lived.

Modern gratings are interference based. That is, they (the master gratings, from which copies are made) are manufactured by the interference pattern of two lasers impinging on a photosensitive surface (thus, the term *holographic*). This creates lines of very specific spacing on a reflective (aluminum, silver, gold) surface. When the polychromatic light from a source strikes the surface, it acts like thousands of prisms, dispersing the light according to Bragg's law:

$$m\lambda = 2d \sin \theta \qquad (2.2)$$

where m is the order of diffraction, λ is the wavelength (of light emitted), d is the distance between lines of the grating, and θ is the angle of diffraction.

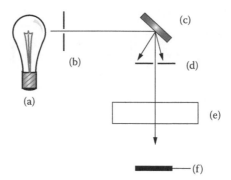

Figure 2.2 A scanning grating-based spectrometer: (a) source, (b) entrance slit, (c) grating, (d) exit slit, (e) sample, and (f) detector.

Depending on the angle at which the grating is rotated, monochromatic light is emitted. The major difference between prisms and gratings is m in the equation. If, for instance, an analyst wishes a wavelength of 2000 nm, the grating is turned to the appropriate angle and the wavelength emitting from the exit slit is 2000 nm. However, also present are wavelengths of 1000 nm, 667 nm, 500 nm, 400 nm, etc. As the order goes from 1 to 2 to 3, the wavelengths add as well. What is needed is an order sorter; that is, a series of cutoff filters is placed in a configuration such that as the grating rotates, the "paddle" moves in concert.

This coordinated effort allows the pure order of light to emerge and impinge on the sample. Some instruments allow all the orders to impinge on the sample, assuming the detectors will ignore the higher-energy orders (visible or UV) and register only the NIR. The analyst must decide whether visible light will negatively impact the sample.

To produce a workable, high signal-to-noise spectrum, the analyst usually co-adds a number of single spectra (32 or 64 scans are typical) to reduce the noise apparent in an individual scan. This means that the typical laboratory grating-based instrument can scan about one sample per minute. This is fast compared with the first scanning spectrometers, but may be too slow for real-time quality control analyses.

2.3   Interferometer-based instruments

Developed in the late 19th century by Michelson, the moving mirror interferometer was initially designed to determine the speed of light. While using the instrument, Michelson noticed some differences in the interference pattern when various materials were placed in the beam. However, it was not until roughly 1960 when the math treatment of Fourier was applied to the interferogram produced by the device that an infrared

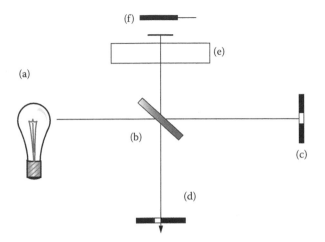

Figure 2.3 An interference-type spectrometer: (a) source, (b) beam splitter, (c) fixed mirror, (d) moving mirror, (e) sample, and (f) detector.

spectrum was seen. The speed and advantages of the Fourier transform infrared (FT-IR) instrument rapidly made it a clear choice over either the prism or grating models for most laboratory applications in the mid- and far-infrared regions.

A simple interferometer is seen in Figure 2.3. The light from the source is split into two segments by the beam splitter.

One portion travels to a fixed mirror and is reflected back to the splitter. The second impinges on a moving mirror and returns to be recombined with the first portion of light. The pattern of peaks and valleys (caused by constructive and destructive interferences, hence the name interferometer) is called an interferogram. When deconvoluted by some mathematical means, e.g., a Fourier transform, a spectrum is obtained.

Any well-behaved periodic function (such as a spectrum) can be represented by a Fourier series of sine and cosine functions of varying amplitudes and harmonically related frequencies. A typical NIR spectrum may be defined mathematically by a series of sines and cosines in the following equation:

$$f(\lambda) = a_0 + (a_1 \sin \omega\lambda + b_1 \cos \omega\lambda)$$

$$+ (a_2 \sin 2\omega\lambda + b2 \cos 2\omega\lambda) \qquad (2.3)$$

$$+ \ldots (a_n \sin n\omega\lambda + b_n \cos n\omega\lambda)$$

where a_0 is the average of the spectrum (mean term), a_i and b_i are the weighting factors (or coefficient pairs), and $n = 1, 2, 3, \ldots$, is the Fourier index.

This equation turns the interferogram into a spectrum or can convert a spectrum into workable pieces for consideration. This Fourier factor concept was explored by McClure [2] as a means of extracting the various components of noise and data from NIR spectra. For instance, the first factor has the largest sine wave. Since the sloping baseline associated with diffuse reflection is caused largely by scattering, this piece of the spectrum could be discarded by reconstructing the spectrum without the first Fourier term. Often, the later Fourier terms describe the instrument noise (high-frequency signals versus broad, low-frequency absorbance bands), and dropping them from the reconstructed spectrum has a smoothing effect.

The popularity of interferometers in mid-range infrared has carried over to NIR. The FT-NIR instruments are becoming quite common in the field. Speed and high-resolution spectra are the strengths of FT instrumentation. Since most spectroscopists are familiar with FT-IR instrumentation, it is logical that they would lean toward an instrument that seems familiar. Another advantage that FT-NIR instruments have over others is the wavelength accuracy. A laser is used to ensure that the *x*-axis is precise and reproducible. This is an advantage that is important when model transfer is part of the implementation strategy.

2.4   Acousto-optic tunable filters

Acousto-optic tunable filters (AOTFs) are solid-state, rapid scanning instruments. The heart of the instrument is an anisotropic crystal (Figure 2.4) capable of splitting a particular wavelength of light into its ordinary and extraordinary (clockwise and counterclockwise rotated) rays. One face (orthogonal to the light beam) has a transducer bonded to it.

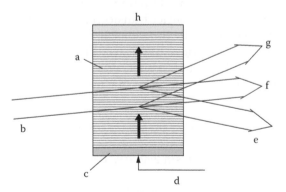

Figure 2.4 An acousto-optic tunable filter: (a) tellurium dioxide (TeO$_2$) crystal, (b) incident or input beam, (c) acoustic transducer, (d) rf input, (e) monochromatic light (ordinary beam), (f) nonscattered light beam, (g) monochromatic light (extraordinary beam), and (h) acoustic wave absorber.

This element converts radio frequencies to sound waves, propagated through the crystal. A dampener is placed on the face opposite to the generator. This effectively reduces returning wavefronts. The result is standing waves throughout the crystal acting as surfaces to refract incoming light. The wavelength of light refracted is based on Bragg's law (Equation (2.2)).

Since radio frequencies can be changed or scanned rapidly, the spacing of the wavefronts can be changed with incredible speed. However quickly the sound waves change spacing, it is still much slower than the speed of light. In effect, the light always sees a standing wave. The overall effect is a rapidly scanning, solid-state monochromator.

The instruments built with AOTFs as monochromators are often connected to fiber optics to obviate problems with polarized light. The instruments are also always designed in such a manner as to stabilize the crystal thermally, as the wavelength emitted per frequency applied is quite temperature sensitive.

In some equipment, one of the monochromatic rays is blocked, allowing the second beam to emulate a typical single-beam instrument. In others, the second ray is used as a reference, allowing the feedback control to keep the output of light at a constant level.

2.5   *Photodiode arrays*

In lieu of a moving monochromator and a single detector, this type of instrument utilizes a fixed monochromator, usually a holographic grating and an array of many small detectors (see Figure 2.5). The light is collimated onto the grating and is in turn dispersed into component wavelengths. These discrete wavelengths are then directed to a series

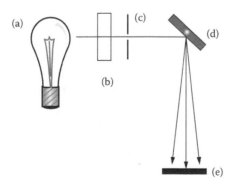

Figure 2.5 A diode array spectrometer: (a) source, (b) sample, (c) slit, (d) fixed monochromator, and (e) detector.

of photodiodes. The array is capable of measuring 64, 128, 256, or more wavelengths simultaneously.

Obviously, the more diodes present in the array, the better the resolution of the spectrum. The cost and complexity of the array and the electronics associated with a greater number of diodes increase with the number of diodes utilized. However, with large, overlapping peaks in the NIR, it may suffice to use merely 256 diodes for satisfactory results.

Laboratory instruments based on photodiode arrays (PDAs) are not commonplace. The trend seems to be applications in process control to avail the analyst of their rapid response.

2.6   Specialty and custom instruments

For several decades, instrument manufacturers have made process instruments by placing the bench instrument they are selling in an explosion-proof casing. Often these were simply the same components as in their laboratory models, merely dissembled and put into industrial casings. A number of nontraditional manufacturers are beginning to design and produce components using the triple credo of smaller, faster, and cheaper. Some of these components are described herein.

2.6.1   Linear variable filters

Since interference filters determine the wavelength passed by the difference in refractive indexes or Fabry-Pérot spacing (see preceding Section 2.1), making the spacer or outer materials differ in refractive index from one end of the filter to the other will produce the effect of a grating. In other words, a spectrum may be generated if the filter is designed properly. One approach to designing these filters is to change the thickness of a bandpass filter coating to let through only the desired wavelengths. The collection electronics and light source can be mounted against the filter to create a compact, low-cost spectrometer.

In recent years, a number of manufacturers, generally from the telecommunication industry, have entered the market, initially as suppliers to other companies, but increasingly targeting directly end customers. The advantage of these linear variable filters (LVFs) resides in the variety of applications that are suited for the small footprint of the system. While originally developed for visible to very near-infrared applications, instruments exist covering the 950 to 1650 nm range and work is in progress to reach longer wavelengths. Spectral collection times are typically measured in microseconds, allowing real-time monitoring of a process. While more work is yet to be done, a number of presentations and publications have shown that performance can be similar to more traditional instrument designs [4–6].

These LVFs are indicative of the solid-state spectrometers expected in the near future. True process control will be possible only when the cost of process instruments drops from its current level. Simple instruments such as these will allow monitoring throughout the process stream.

2.6.2 *Miniature photodiode arrays*

Several very small photodiode arrays are commercially available. The first commercial units were built by Ocean Optics (Dunedin, Florida), but they are now available from several sources. These devices are based on the light (postsample, either diffusely reflected or transmitted) impinging upon a very small fixed grating, then onto a series of diodes with minimum spacers between them. While the number of diodes and resolution of the array are limited, the small size and high speed of these types of instruments may compensate for the lack of resolution in a variety of applications and may provide certain advantages in the appropriate setting.

As with the linear variable filters, these mini-monochromators will likely revolutionize pharmaceutical production. Small, rugged, and inexpensive devices are the wave of the future; measurements will not only be taken more often, but at many more points along the process stream.

2.6.3 *Single-wavelength attenuated total reflection-based devices*

Wilks Enterprises (Norwalk, Connecticut) [3] introduced a device incorporating a miniature attenuated total reflection (ATR) cell as its interface. The device resembles an old-fashioned electrical fuse for a home fuse box. In place of the metal contact at the base, a small ATR cell protrudes from the base. Monochromatic light courses through this.

The light is produced by a small, single-wavelength source within the housing and detected by a small detector alongside it. Nothing extends from the housing but several wires carrying power to and the signal from the device. While the current models were devised for mid-range IR, later models will encompass the NIR.

The importance of these bullet detectors is that a cluster of them might give similar data at a fraction of the cost when compared to a monochromator-based instrument. In fact, as the cost declines for these devices, it could (conceivably) be less expensive to discard older sensors than to recalibrate them. (Disposable instruments? Why not? We already have disposable cameras and cell phones!)

2.6.4 *Multivariate optical computing*

Multivariate optical computing (MOC) is a very recent and innovative strategy to build near-infrared spectrometers. It is based on the creation of multivariate optical elements (MOEs) that can be installed into a spectrometer to estimate the value of a parameter of interest. In a nutshell, MOC is analogous to the dot product of absorption and regression parameters that a traditional regression model would follow: the regression vector is encoded in the MOE. Using MOC, an interference pattern is calculated and encoded into an optical filter to allow the estimation of a parameter of interest. The MOE is then created following a process similar to how interference filters are produced.

Multivariate optical elements make prediction in a way similar to that of principal component regression. The light coming from a sample is transmitted and reflected by the MOE. The difference between the transmitted and reflected light is proportional to the chemical or physical property of interest (for which the MOE was designed). As stated in the original article presenting the technology, "Using simple optical computing, multivariate optical elements mimic the performance of a full spectroscopy system using traditional analysis. The throughput and efficiency of the system is two to four orders of magnitude higher than conventional spectrometers, making lower-cost detectors practical" [7].

Multivariate optical elements provide a number of advantages over other types of spectrometers: the cost of creating the filters is low, and the design of the spectrophotometer is simple; there is no moving part, thus reducing the need for instrument maintenance, and the instrument footprint is limited, with devices that can be reduced to hold in the hand.

While not all specific to near infrared, a number of publications have shown the applicability of multivariate optical elements [8–11]. It is anticipated that the technology will improve in the coming years, and that the number of applications will increase. A review of the current applications was performed by Priore [12].

2.6.5   Micro-electro-mechanical system

Micro-electro-mechanical system (MEMS)-based spectrometers are a new type of instrumentation design. The main driver behind the development of MEMS technology is to provide a rugged online spectrometer with a very limited footprint. The cost is also at the core of this disruptive technology. Micro-electro-mechanical systems use the technology developed by the semiconductor industry to manufacture spectrometer components. Using lithographic techniques, a large number of elements can be manufactured on a wafer at the same time, thus driving down the cost of each spectrometer. The MEMS development is rooted in the boom of the telecommunication industry. A special class of MEMS is MEOMS (micro-opto-electro-mechanical systemss), which senses and manipulates light.

A micro-electro-mechanical system combines sensors, actuators, and electronics on a single closed-loop system. A super luminescent light-emitting diode is typically used as a light source, and light diffraction is achieved by either a miniature Fabry-Pérot filter or a grating. Other MEMS designs exist, such as the Hadamard transform spectrometer or the Fourier transform spectrometer. For more information about MEMS, readers should refer to a very comprehensive review by Schuler et al. [13] and a series of articles by Crocombe [14–17].

Schuler et al. [13] noted in his review the following: "Large optics and detector arrays are replaced with high-speed MEMS-tunable filters and gratings. Preferably, a single, wide bandwidth detector is used, that enables the manufacture of extremely rugged spectrometers without many of the alignment issues and at lower cost. However, with miniaturized devices the optical path lengths are reduced which generally leads to decreased optical performance [18]. At present, most, if not all MEMS-based spectrometers have a lower performance than the best conventional instruments." Nevertheless, the fit-for-purpose mechanism must apply when deciding if it is more appropriate to invest in an expensive bench-top unit and attempt to apply it to processes, or if a rugged, small, and cost-efficient system can provide the necessary information for process control. This is the same argument presented by LVF and other specialty instrument manufacturers.

2.7 Optical parameter instrumentation

The separation of absorption and scattering often requires specific types of instrumentation, based on the theoretical approach that is used. In this section, four main designs will be discussed: integrating sphere-based reflectance and transmittance, spatially resolved spectroscopy, time-resolved spectroscopy, and frequency domain photon migration.

Using one or two integrating spheres, it is possible to isolate reflected and transmitted light. With one integrating sphere, the reflection and transmission signals are measured by changing the location of the sample on the sphere. For reflection, the incident light source interacts with the sample after entering the sphere while in transmission mode, the incident light hits the sample, and only the fraction transmitted enters the sphere. A more practical setup uses two spheres, and the sample is at the interface. The first sphere will collect the reflected light, while the second will measure the transmitted signal. The derivation of the absorption and scattering coefficients from the measured reflected and transmitted light can be found in Kuhn et al. [19]. Pharmaceutical applications for this design were published by Burger et al. [20, 21].

Spatially resolved spectroscopy consists in measuring the light reflected at different distances from the incident light. Typical designs use a set of aligned optical fibers, with one fiber being the source and others the detection systems. Setups based on hyperspectral imaging have also been proposed [22]. Absorption and scattering coefficients may be calculated from the change in the signal as a function of the distance from the incident light. The derivation of the optical parameters from the measured reflected light can be found in Farrell and Patterson [23]. Pharmaceutical applications of this equipment design were published by Shi and Anderson [22, 24] for pharmaceuticals and Kienle et al. [25] for medical applications.

Another approach to the calculation of optical parameters is based on time-resolved spectroscopy. This method relies on the temporal dispersion of a light pulse (few picoseconds) as it propagates through a turbid medium. The light remitted by the sample at a given distance from the source is recorded. The temporal shape of the light pulse is modified as it passes through the sample, and based on the modification of the shape, the optical properties of the sample can be estimated. The derivation of the absorption and scattering coefficients from the temporal shapes can be found in Patterson et al. [26]. Pharmaceutical applications of this design were published by Abrahamsson et al. [27] and Johansson et al. [28].

The last design, representing frequency domain photon migration, is based on the change in light modulation by the sample. The modulation of the light after interacting with the sample is compared with the incident modulation. The measurements of phase-shift and amplitude attenuation can be determined as a function of optical properties of the sample medium. The derivation of the absorption and scattering coefficients from the modification of the modulation can be found in Sun et al. [29]. The research group of the same author published several articles on the use of frequency domain photon migration in pharmaceutical products [30–33] .

While these instrument designs may appear most suitable for laboratory analysis, some manufacturers have proposed apparatus suitable for real-time online process monitoring. It is the case for spatially resolved spectroscopy. A very good review on the topic on pharmaceutical applications to the separation of absorption and scattering was written by Shi and Anderson [34]. A comprehensive review of those techniques and applications in tissue optics can be found in a review by Wilson [35].

2.8   Strengths and shortcomings of traditional types of equipment

Every type of instrument has some inherent advantage for any particular analysis. For example, water is the most analyzed material in NIR. If this is the only analysis to be run for a particular product, it may well be sufficient to use a simple filter instrument. Using a $60,000 to $100,000 instrument to measure only moisture is a waste of resources.

This does not mean that a company might not wish to purchase the most versatile instrument in every case, anticipating more applications in the future. Nor does it mean that only certain instruments will only work for certain applications. It only means that an equation built on an optimal instrument for a task has a higher percentage chance of success.

2.8.1   Filter instruments

The most rugged instruments available, filter-based devices are capable of performing rather sophisticated analyses. With a filter wheel providing several dozen wavelengths of observation, multiple-analyte analyses are quite common. These multiple assays are more amenable to agricultural or food products than pharmaceuticals, if only for validation purposes. Fine chemicals, pharmaceuticals, and gases may be likely to have sharper peaks and are better analyzed via a continuous monochromator (i.e., grating, FT, AOTF).

The strength, then, of filters is the ruggedness of design. The weakness lies in the broad band allowed through the filter and lack of full spectral signature. For simple analyses involving major components, the resolution of filters is sufficient. For fine work, a more sophisticated monochromator may be needed.

2.8.2   Acousto-optic tunable filter instruments

In general, AOTF-based instruments are rugged and fast, capable of numerous readings per second. As long as there is no radio frequency interference, the wavelength accuracy and reproducibility are excellent. While individual instruments vary, the nominal spectral resolution is around 10 nm. This is sufficient for process purposes, but if greater resolution is needed, a FT-NIR is required.

The light beam emitted by the AOTF crystal is circularly polarized and may be run through a fiber optic before striking a sample to minimize polarization effects. In addition, since the energy associated with any particular wavelength is split into two rays, the light reaching a sample from any particular source will be approximately one-half that produced by a grating-based instrument. For some applications, a stronger source may be required.

The in-the-field history of these instruments is relatively short, and the service record that is available for grating and FT instruments is still to come. However, the durability of these devices with no moving parts appears to be excellent. Additionally, original concerns over crystal uniqueness causing calibration transfer problems appear to be insignificant.

2.8.3   Grating instruments

Still the workhorse of the analytical lab, grating instruments may be somewhat slow for modern process applications. Since most classical equations have been generated on grating instruments, there is a wealth of information available for the grating user. The better instruments are quite low in noise and have excellent reproducibility.

However, on the negative side, the moving grating can be knocked out of calibration by vibrations in a process environment. Indeed, any moving part will, by definition, wear out in time.

2.8.4   Fourier transform–NIR instruments

These are extremely fast with the highest possible resolution, although the scan averaging required for adequate signal-to-noise may provide ultimate scan times that are similar to those of other instrumentation. They are the best bet for analyzing gases or materials where resolution is critical (due to a sharp λ_{max}). Many active pharmaceutical ingredients (APIs) are crystalline and have sharp peaks, even in the NIR. In such cases, the resolution of a FT instrument is a plus.

The usual advantages of FT-IRs do not necessarily apply in the NIR, however. The throughput and multiplex advantages (of mid-range IR) only apply where the source is weak, absorbencies are strong, and the detector is not too sensitive.

In this scenario (mid-range infrared), the extra energy is needed to (1) penetrate the sample and (2) raise the signal-to-noise (S/N) ratio to a level where small peaks are visible and quantitation is possible.

With NIR, the source is quite strong, the absorbencies weak, and the detector(s) quite quiet and sensitive. The extra energy can, in some cases, burn the sample. Lactose is an example of a pharmaceutical excipient browned by the energy of FT-NIR. In addition, since most NIR peaks are broad and relatively featureless, high resolution usually adds noise (which is, by nature, high frequency).

2.8.5   Photodiode arrays

Virtually instantaneous and rugged, PDAs have yet to be fully utilized in process control. Larger models are difficult to use for production, as they were designed for static samples. They also require collection of the light diffusely reflected from the solid samples via mirrors and lenses. Most commercial designs are excellent for lab use, but not rugged enough for the production line.

2.9 Summary

A large variety of suitable NIR equipment is available for pharmaceutical applications. It is up to the analyst to ascertain the correct instrument(s) for the particular application. Determination of the proper equipment can only be achieved by first understanding the chemistry/physics of the process. After the key measurements have been determined, the correct tool for the job may be chosen.

Instrument company applications personnel should be consulted early in the method development process for assistance with the identification of suitable equipment. It is also a good idea to research the repair track record of the company being considering as a supplier. An instrument is worthless if not working nearly 100% of the time. All instruments break down. How quickly and how well an instrument company responds to a breakdown should be a strong determinant in whether an instrument is worth purchasing. Finally, and possibly most importantly, the instrument vendor must be willing and able to support your method validation efforts.

References

1. E. W. Ciurczak, Selecting a Spectroscopic Method by Industrial Application, in *Applied Spectroscopy: A Compact Reference for Practitioners*, ed. J. Workman and A. Springsteen, Academic Press, New York, 1998, p. 329.
2. W. F. McClure, Analysis Using Fourier Transforms, in *Handbook of Near-Infrared Analysis*, ed. D. A. Burns and E. W. Ciurczak, Marcel Dekker, New York, 1992, p. 181.
3. E. W. Ciurczak, A New In-Line Infrared Sensor, *Spectroscopy*, 16(1), 16 (2001).
4. B. Igne, M. N. Hossain, C. A. Anderson, and J. K. Drennen, Near Infrared Calibration Life-Cycle Management of the Active Content of an Oral Dosage Form, presented at SCIX 2013, Milwaukee.
5. C. Pederson, Real-Time Tablet API Analysis: A Comparison of a Palm-Size NIR Spectrometer to HPLC Method, presented at IFPAC 2014, Washington, DC.
6. M. Alcalà, M. Blanco, D. Moyano, N. W. Broad, N. O'Brien, D. Friedrich, F. Pfeifer, and H. W. Siesler, Qualitative and Quantitative Pharmaceutical Analysis with a Novel Handheld Miniature Near-Infrared Spectrometer, *J. Near Infrared Spectrosc.*, 21, 445 (2013).
7. M. P. Nelson, J. F. Aust, J. A. Dobrowolski, P. G. Verly, and M. L. Myrick, Multivariate Optical Computation for Predictive Spectroscopy, *Anal. Chem.* 70, 73 (1998).
8. C. M. Jones, R. Atkinson, D. Chen, M. Pelletier, D. Perkins, J. Shen, and B. Freese, Laboratory Quality Optical Analysis in Harsh Environments, presented at SPE Kuwait International Petroleum Conference and Exhibition, Kuwait City, Kuwait, December 10–12, 2012, SPE-163289-MS.

9. J. A. Swanstrom, L. S. Bruckman, M. R. Pearl, M. N. Simcock, K. A. Donaldson, T. L. Richardson, T. J. Shaw, and M. L. Myrick, Taxonomic Classification of Phytoplankton with Multivariate Optical Computing. Part I. Design and Theoretical Performance of Multivariate Optical Elements, *Appl. Spectrosc.*, 67, 620 (2013).

10. J. A. Swanstrom, L. S. Bruckman, M. R. Pearl, E. Abernathy, M. N. Simcock, K. A. Donaldson, T. L. Richardson, T. J. Shaw, and M. L. Myrick, Taxonomic Classification of Phytoplankton with Multivariate Optical Computing. Part II. Design and Experimental Protocol of a Shipboard Fluorescence Imaging Photometer, *Appl. Spectrosc.*, 67, 630 (2013).

11. M. R. Pearl, J. A. Swanstrom, L. S. Bruckman, T. L. Richardson, T. J. Shaw, H. M. Sosik, and M. L. Myrick, Taxonomic Classification of Phytoplankton with Multivariate Optical Computing. Part III. Demonstration, *Appl. Spectrosc.*, 67, 640 (2013).

12. R. Priore, Multivariate Optical Elements Beat Bandpass Filters in Fluorescence Analysis, *J. Laser Focus World*, 49, 49 (2013).

13. L. P. Schuler, J. S. Milne, J. M. Dell, and L. Faraone, MEMS-Based Microspectrometer Technologies for NIR and MIR Wavelengths, *J. Phys. D Appl. Phys.*, 42, 133001 (2009).

14. R. A. Crocombe, Miniature Optical Spectrometers: There's Plenty of Room at the Bottom. I. Background and Mid-Infrared Spectrometers, *Spectroscopy*, 23, 38 (2008).

15. R. A. Crocombe, Miniature Optical Spectrometers: Follow the Money. II. The Telecommunications Boom, *Spectroscopy*, 23(3 Suppl), 25 (2008).

16. R. A. Crocombe, Miniature Optical Spectrometers. III. Conventional and Laboratory Near-Infrared Spectrometers, *Spectroscopy*, 23, 40 (2008).

17. R. A. Crocombe, Miniature Optical Spectrometers: The Art of the Possible. IV. New Near-Infrared Technologies and Spectrometers, *Spectroscopy*, 23, 26 (2008).

18. R. F. Wolffenbuttel, State-of-the-Art in Integrated Optical Microspectrometers. Instrumentation and Measurement, *IEEE Trans.*, 53, 197 (2004).

19. J. Kuhn, S. Korder, M. C. Arduini-Schuster, R. Caps, and J. Fricke, Infrared-Optical Transmission and Reflection Measurements on Loose Powders, *Rev. Sci. Instrum.*, 64(9), 2523 (1993).

20. T. Burger, J. Kuhn, R. Caps, and J. F. Fricke, Quantitative Determination of the Scattering and Absorption Coefficients from Diffuse Reflectance and Transmittance Measurements: Application to Pharmaceutical Powders, *Appl. Spectrosc.*, 51(3), 309 (1997).

21. T. Burger, H. J. Ploss, S. Ebel, and J. Fricke, Diffuse Reflectance and Transmittance Spectroscopy for the Quantitative Determination of Scattering and Absorption Coefficients in Quantitative Powder Analysis, *Appl. Spectrosc.*, 51(9), 1323 (1997).

22. Z. Shi and C. A. Anderson, Application of Monte Carlo Simulation-Based Photon Migration for Enhanced Understanding of Near-Infrared (NIR) Diffuse Reflectance. Part I. Depth of Penetration in Pharmaceutical Materials, *J. Pharm. Sci.*, 99, 2399 (2010).

23. J. Farrell and M. S. Patterson, A Diffusion Theory Model of Spatially Resolved, Steady-State Diffuse Reflectance for the Noninvasive Determination of Tissue Optical Properties In Vivo, *Med. Phys.*, 19(4), 879 (1992).

24. Z. Shi and C. A. Anderson, Application of Monte Carlo Simulation-Based Photon Migration for Enhanced Understanding of Near-Infrared (NIR) Diffuse Reflectance. Part II. Photon Radial Diffusion in NIR Chemical Images, *J. Pharm. Sci.*, 99, 4174 (2010).

25. A. Kienle, L. Lilge, M. S. Patterson, R. Hibst, R. Steiner, and B. C. Wilson, Spatially Resolved Absolute Diffuse Reflectance Measurements for Noninvasive Determination of the Optical Scattering and Absorption Coefficients of Biological Tissue, *Appl. Opt.*, 35(13), 2304 (1996).

26. M. S. Patterson, B. Chance, and B. C. Wilson, Time Resolved Reflectance and Transmittance for the Noninvasive Measurement of Tissue Optical Properties, *Appl. Opt.*, 28(12), 2331 (1989).

27. C. Abrahamsson, A. Lowgren, B. Stromdahl, T. Svesson, S. Andersson-Engels, J. Johansson, and S. Folestad, Scatter Correction of Transmission Near-Infrared Spectra by Photon Migration Data: Quantitative Analysis of Solids, *Appl. Spectrosc.*, 59(11), 1381 (2005).

28. J. Johansson, S. Folestad, M. Josefson, A. Sparen, C. Abrahamsson, S. Andersson-Engels, and S. Svanberg, Time-Resolved NIR/Vis Spectroscopy for Analysis for Solids: Pharmaceutical Tablets, *Appl. Spectrosc.*, 56(6), 725 (2002).

29. Z. G. Sun, Y. Q. Huang, and E. M. Sevick-Muraca, Precise Analysis of Frequency Domain Photon Migration Measurement for Characterization of Concentrated Colloidal Suspensions, *Rev. Sci. Instrum.*, 73(2), 383 (2002).

30. T. Pan, D. Barber, D. Coffin-Beach, Z. Sun, and E. M. Sevick-Muraca, Measurement of Low-Dose Active Pharmaceutical Ingredient in a Pharmaceutical Blend Using Frequency Domain Photon Migration, *J. Pharm. Sci.*, 93(3), 635 (2004).

31. T. Pan and E. M. Sevick-Muraca, Volume of Pharmaceutical Powders Probed by Frequency-Domain Photon Migration Measurements of Multiply Scattered Light, *Anal. Chem.*, 74(16), 4228 (2002).

32. R. R. Shinde, G. V. Balgi, S. L. Nail, and E. M. Sevick-Muraca, Frequency-Domain Photon Migration Measurements for Quantitative Assessment of Powder Absorbance: A Novel Sensor of Blend Homogeneity, *J. Pharm. Sci.*, 88(10), 959 (1999).

33. E. M. Sevick-Muraca, Z. G. Sun, and T. S. Pan, Characterizing Powders Using Frequency-Domain Photon Migration, U.S. Patent 6771370 (2004).

34. Z. Shi and C. A. Anderson, Pharmaceutical Applications of Separation of Absorption and Scattering in Near-Infrared Spectroscopy (NIRS), *J. Pharm. Sci.*, 99, 4766 (2010).

35. B. C. Wilson, Measurement of Tissue Optical Properties: Methods and Theories, in *Optical-Thermal Response of Laser-Irradiated Tissue*, ed. A. J. Welch and M. J. C. van Gemert, Plenum Press, New York, 1995, p. 233.

chapter 3

Blend uniformity analysis

3.1 Mixing

Mixing is a basic pharmaceutical processing operation required for the production of most pharmaceutical products. Liquid, solid, or semisolid products are mixed in a large variety of mixers, blenders, and milling machines. The ultimate goal of any mixing process is to achieve content uniformity, a situation where the contents of a mixture are uniformly distributed. Such a mixture is said to be homogenous. An exact definition of homogeneity is elusive, however, because the large varieties of pharmaceutical substances that are mixed show a multiplicity of mixing mechanisms, and the nature of the final optimal blends of such mixtures are so diverse. While this chapter is not intended as a review of mixing, some of the practical concerns for any mixing process will be discussed in relation to the possible advantages of using near-infrared (NIR) methods for monitoring or controlling blending.

The determination of powder blend homogeneity is generally more complex than homogeneity determinations for other blend types due to the many factors that complicate powder blending, such as particle size, shape, density, resilience, and surface roughness. Also, the humidity of the local environment and electrostatic charge on the surface of materials to be blended is important. Mixing and segregation processes compete as powders are blended. Optimal blend times depend on the actual materials involved and the efficiency of the particular blender that is used. A thorough discussion of ordered and random powder mixtures and the various approaches to identifying process endpoints could easily fill a tome and is not appropriate for this chapter. The reader should keep in mind that a NIR spectrum provides chemical and physical information about all components of a sample. Such chemical and physical data should be valuable for homogeneity determination of all blend types.

3.2 Discussion of reported work

Because most pharmaceutical active ingredients and excipients absorb NIR radiation, studies utilizing NIR may provide certain advantages over traditional methods of assay for blend homogeneity or complement the traditional assay for active ingredient by providing homogeneity information regarding all mixture components. Additionally, a direct non-destructive method for assessing blend homogeneity could be of great value in minimizing the sample preparation and assay time associated with traditional blend analysis procedures. The fact that NIR spectroscopic methods can be used noninvasively may even eliminate the need for using a sample thief to remove samples from a mixer, thereby reducing the error associated with such sampling techniques.

Drennen and Lodder were among the first researchers to use NIR for monitoring pharmaceutical mixing [1, 2]. Their work involved the analysis of a petrolatum-based ointment to determine when homogeneity had been achieved. The authors commented on several possible approaches to the spectral measurements needed for NIR determination of homogeneity of any pharmaceutical blend. As any mixing process proceeds, multiple spectra collected from the mixture should approach the spectrum that represents a homogenous blend. The multiple spectra may be collected in two ways, in either time or space. In the first case, the spectra may come from one point in the mixing vessel, say, through a single window over a period of time as the mixing continues. The achievement of homogeneity may be identified by one of three means in such studies:

1. The individual spectrum may be compared to a library of spectra that arise from previous homogenous blends. In this case, content uniformity is apparent when the spectrum or spectra collected from the current mixing is deemed similar to the library of spectra that represent previous homogenous blends. Any suitable algorithm for qualitative spectral identification would be appropriate for such calculations.
2. The spectra from the current blend, collected at different time points during the mixing process, may be compared to each other. Again, suitable algorithms for spectral comparison may be used, or more common statistical parameters such as a standard deviation or chi-square, in order to identify similarity of spectra. Using such common parameters as standard deviation or a test of homogeneity of variance (as in Bartlett's chi-square test) allows the spectroscopist to use individual absorbance values at wavelengths where the active ingredient or other significant components have absorbance bands. Statistical decisions regarding the endpoint of blending can be made with whatever significance level is required for a given application. For example, mixing of antibiotic with a veterinary feed product may not require the same significance as a high-potency heart medication.
3. Quantitative methods may be used to predict the homogeneity of the blend for the parameter(s) of interest based on the spectra. This approach varies from the quantitative analysis of spectra due to the fact that the regression model used to output predictions might

emphasize certain absorption regions more than others for a particular chemical species. This is particularly valid when the control of the homogeneity of several ingredients is desired.

The second and third means of collecting multiple spectra involve collection from numerous locations within the vessel at one or more times. Multiple spectra drawn at one time from various locations in a blender may be compared to themselves or to spectra collected at different times, with similarity indicating content uniformity. Parameters for comparison can again include common statistical factors such as standard deviation, relative standard deviation, variance, a host of indexes based on standard deviation, or the results from pattern recognition routines. Of course, spectra collected in this manner can also be compared to a library of spectra from previous homogenous blends.

The use of NIR for homogeneity determination does not necessarily require the development of a calibration model. The previously described method could be applied to any mixture without the extensive preliminary work of building a quantitative calibration as required for most NIR work. On the other hand, homogeneity determinations could be carried out by using either quantitative or qualitative calibrations. Quantitative calibrations involve prediction of the drug level at various locations or times with an appropriate calibration. The standard deviation of multiple potency determinations may be used to estimate homogeneity.

Regardless of the type of formulation being blended, a noninvasive and nondestructive NIR spectral measurement can simplify homogeneity determinations. Homogeneity determinations are most difficult for blended powders. Numerous factors affect the correctness of homogeneity determinations, including the nature of the mixture, the method of analysis, and sampling methods. Regarding sampling methods, the location of sampling, the number of samples drawn, the size of individual samples, and even the design of the sample thief are important considerations. Many authors have discussed the problems associated with drawing samples from a powder bed using a thief. Poux and Fayoulle [3] have summarized some of the problems identified by other researchers involving thief probes drawing samples that are not representative of the mixture.

Notwithstanding the difficulties, preparation of a uniform powder blend prior to tabletting or encapsulation is a vital step in the production of solid pharmaceutical dosage forms. The determination of powder blend homogeneity is typically a labor-intensive process involving the removal of unit-dose samples from defined mixer locations using a sample thief, extraction of the active drug from the sample matrix, and drug content analysis by either high-performance liquid chromatography or direct UV spectroscopy. The distribution of individual excipients is typically assumed to be homogenous if the active ingredient is uniformly distributed. The idea of using optic probes to monitor blending of powders is not new. As early as 1957, reports of reflection measurements for the purpose of measuring mixture composition were reported [4].

In 1966, Ashton et al. [5] likewise reported the use of a light probe for assessing the homogeneity of powder mixtures. Ashton et al. used a reflection probe to collect at least 50 measurements per time point from a 1 ft^3 blender during mixing. Calibrating with five or six mixtures of known composition, the output from a photocell was then used to assess the degree of "mixedness" during the mixing operation. The authors also discussed the fact that the determination of content uniformity is dependent on the sample size. It is therefore important to recognize the sample size that is equivalent to the sample measured by the light probe. The sample size corresponding to the variance of results obtained with the probe was found from a plot of variance against sample size based upon manual sampling of the mixture using samplers of different capacities. Equivalent sample weights varied depending on the particular mixture under study.

3.2.1 Qualitative approaches

Drennen and Lodder used a bootstrapping pattern recognition algorithm to compare sets of spectra, collected noninvasively from various times during the mixing of an ointment [1, 2]. At early times during mixing, before homogeneity was achieved, comparison of spectral sets revealed that spectra were not similar. After homogeneity was achieved, however, spectral sets collected at different times were identical, within the predefined statistical limits.

Ciurczak reported the use of NIR spectroscopy in powder mixing studies, discussing his use of a spectral matching routine and principal component analysis (PCA) for distinguishing spectra arising from samples drawn at various times during mixing [6]. Principal component analysis proved to be the most sensitive algorithm for discerning final mixtures from penultimate mixtures.

After collecting spectral data at different locations in the blender and at various time points, Sanchez et al. [7] evaluated several qualitative blend endpoint methods. A mean standard deviation for the spectra collected at each time point was compared over time to determine homogeneity. This method allowed authors to determine the characteristics of the mixture spectrum, corresponding to the desired blend state. Using that mixture spectrum, they defined a dissimilarity index based on the orthogonal projection of spectra collected at various times points onto the mixture spectrum. The norm of the resulting vector was then employed to evaluate the blend state.

An alternative approach used PCA to compare, in score space, the Euclidean distance between spectra and the mixture spectrum. Finally, the self-modeling SIMPLISMA [8] approach was tested in an effort to qualify blends. SIMPLISMA aims at selecting wavelengths with the highest purity (selectivity) with respect to the parameter of interest. "Assuming that Beer's law is followed, each pure wavelength is proportional to the concentration of the compound absorbing." The evolution of the variance of the pure wavelengths was then used to estimate the evolution of homogeneity. Authors determined that PCA provided more insights into the nature of the heterogeneity than the dissimilarity index. SIMPLISMA was found to confirm PCA results.

Sekulic et al. reported the online monitoring of powder blend homogeneity by NIR spectroscopy in 1996 [9]. The authors commented on the error associated with insertion of a sample thief into a powder bed due to distortion of a powder bed, adhesion of certain components to the thief with subsequent displacement, and preferential flow of some components into the thief, all factors that make the NIR method attractive and which are discussed more thoroughly elsewhere. For these reasons and others, the authors developed a method of monitoring the blending process that eliminates the need for withdrawing samples from a blender.

The investigators synchronized the spectral acquisition to the rotational position of the blender and collected spectra through a diffuse reflection probe located on the axis of rotation. Assuming that the fill level is adequate, the probe extends through the axle of the blender into the powder mixture that is being blended. A moving block standard deviation calculation was used to identify the time at which mixing was complete. Although the standard deviations drop to a minimum at the point that homogeneity is reached, as seen in plots of standard deviation versus time, no statistical test was suggested for the identification of homogeneity with any particular degree of certainty.

In a later publication [10], the same group tested the influence of spectral pretreatment techniques on the determination of the endpoint and improved the blend qualification by PCA using soft independent modeling of class analogy (SIMCA). This method is based on the development of independent PCA models for independent groups. Authors showed that by developing a PCA model by only selecting spectra corresponding to homogeneous powder, the prediction of a similar endpoint for additional runs of the same composition was possible. In that publication, authors used sample residual variance and an F-test to estimate if a new sample belongs to the target class. In another manuscript, they used Hotelling's T^2 confidence interval to determine the membership of a spectrum to the desired homogeneity state [11].

Wargo and Drennen discussed their NIR characterization of pharmaceutical powder blends in a 1996 paper [12]. Near-infrared spectroscopy was used to qualitatively assess the homogeneity of a typical direct compression pharmaceutical powder blend consisting of hydrochlorothiazide, fast-flo lactose, croscarmellose sodium, and magnesium stearate. Near-infrared diffuse reflection spectra were collected from thieved powder samples using a grating-based spectrometer. A second-derivative calculation and principal component analysis were performed on the spectra prior to qualitative evaluation. Blend homogeneity was determined using single- and multiple-sample bootstrap algorithms [13, 14] and traditional analysis of variance (ANOVA) using a chi-square test [15] for homogeneity of variance. The results suggested that bootstrap techniques provided greater sensitivity for assessing blend homogeneity than chi-square calculations.

The bootstrap error-adjusted single-sample technique (BEAST) [13] represents a type of analytical procedure designed to operate in the high-speed parallel or vector mode required of pattern recognition tests involving thousands of samples. Lodder and Hieftje have discussed this technique, derived from Efron's bootstrap calculation [16], in detail and have provided examples of its application. The bootstrap calculation of Lodder and Hieftje can be used to provide both quantitative and qualitative analyses of intact products. The BEAST begins by treating each wavelength in a spectrum as a single point in multidimensional space (hyperspace). Each point is translated from the origin along each axis by an amount that corresponds to the magnitude of the signal observed at each wavelength. Samples with similar spectra map into clusters of points in similar regions of hyperspace, with larger cluster size corresponding to samples with greater intrinsic variability.

The BEAST develops an estimate of the total sample population using a small set of known samples. A point estimate of the center of this known population is also calculated. When a new sample is analyzed, its spectrum is projected into the same hyperspace as the known samples. A vector is then formed in hyperspace to connect the center of the population estimate to the new sample spectral point. A hypercylinder is formed about this vector to contain a number of estimated population spectral points. The density of these points in both directions along the central axis of the hypercylinder is used to construct an asymmetric nonparametric confidence interval. The use of a central 68% confidence interval produces bootstrap distances analogous to standard deviations.

Bootstrap distances can be used to identify homogenous blends. Bootstrap distances are calculated as follows:

$$\frac{\sqrt{\left(\sum_{j=1}^{d}\left(c_j - x_j\right)^2\right)}}{\sigma} \tag{3.1}$$

where c_j is the center of the bootstrap distribution, x_j is the test sample spectrum, and σ is a BEAST standard deviation.

When a sample spectrum projects to a point within three standard deviations of the center of a cluster of spectral points from a known substance or product, the sample is considered to be a sample of the known material. When the new sample contains components in concentrations that differ from the well-blended product, the new sample spectral point is displaced from the known spectral cluster. The magnitude of this displacement increases as the difference between the new sample and the set of known samples increases. Furthermore, the direction of the displacement of the new sample point corresponds to the spectra of the constituents responsible for the displacement.

Where the single-sample bootstrap algorithm provides for the qualitative analysis of a single test sample, the modified bootstrap algorithm [14] provides a test that uses multiple test spectra to detect false samples (as subclusters) well within the three-standard deviation limit of a training set. The accurate detection of subclusters allows the determination of very small changes in component concentration or physical attributes.

The first steps of the modified bootstrap test are the same as in the single-sample bootstrap test. A training set is constructed from known samples and a Monte Carlo approximation to the bootstrap distribution is calculated to estimate the population from which the training set is drawn. The center of the cluster represents the best estimate of the spectrum of the compound. The modified bootstrap varies from the single-sample bootstrap by next projecting the bootstrap-estimated test population into the same multidimensional space as the training set and calculating cumulative distribution functions (CDFs) for the estimated populations (training set and test set). As the number of observations (bootstrap replicates) increases, the rough empirical cumulative distribution function (ECDF) approaches the smooth theoretical cumulative distribution function (TCDF).

Plotting the ECDF versus TCDF, for a given probability, generates the linear version of a cumulative distribution function, a quantile-quantile (QQ) plot. Each cumulative probability value yields a pair of order statistics (one from each CDF) that form a point in the QQ plot. Quantile-quantile plots are valuable tools for distinguishing differences in shape, size, and location between spectral clusters. Two similar clusters of spectra will demonstrate a linear QQ plot (Figure 3.1). Breaks or curves in the QQ plot indicate that differences exist between the groups (Figure 3.2).

Spectral similarity is based on the linearity of the quantile-quantile plots generated. A 98% confidence limit for the correlation coefficient of the training set is calculated. Quantile-quantile plots with correlation coefficients less than the calculated confidence limit are considered to be spectrally different than the training group, while QQ plots with

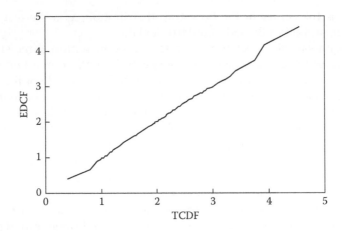

Figure 3.1 A QQ plot representing the situation where the spectral training set and test set are similar.

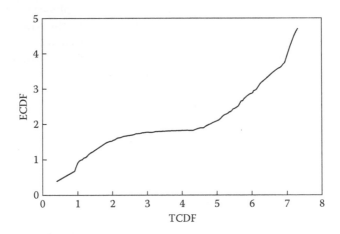

Figure 3.2 A QQ plot representing the situation where the spectral training set and test set are unique.

correlation coefficients greater than the confidence limit are considered spectrally similar. In Wargo's study, five blends of identical composition were subjected to different mixing times, 1, 5, 10, 15, and 20 min, respectively. At the specified time, the blender was stopped and samples equivalent to one to three dosage units, 200–600 mg, were removed from 10 different powder bed locations using a sample thief. In the second part of this study, a single formulation was mixed for 30 min, during which time six unit-dose samples were thieved from the blender at 2 min time intervals. In both experiments, powder samples were transferred to tarred borosilicate sample vials and the sample weights were recorded for reference determination of drug concentration.

Triplicate NIR scans were averaged for each sample. The triplicate spectra were averaged to provide one spectrum per sample and subjected to a second-derivative correction for removal of baseline shifting. Principal component analysis of the second-derivative spectra was used prior to one of two bootstrap calculations, the single-sample or multiple-sample bootstrap calculation. The detection of subclusters by this method allows for the determination of very small changes in component concentrations.

In a third method of evaluating the data, a chi-square analysis was used to assess NIR spectral variability. For each time point, the pooled variance of the NIR absorbance values at individual wavelengths is calculated as the weighted average of the variances. A chi-square statistic is then calculated and compared to a tabulated value for significance at the 5% level, with significant values indicating that the variances are not equal.

During the first multiple-blend study, the multiple-sample bootstrap method offered the greatest sensitivity for certain identification of content uniformity reached at the endpoint of the mixing process. This part of the study also identified a significant advantage of the pattern recognition methods in homogeneity testing over any method that involves a simple study of spectral variance, such as the study of standard deviations at any or all wavelengths. The advantage is that the pattern recognition methods allow not only the recognition of spectral variance, but also identification of errant potencies. The importance can be recognized in the following example. Consider for a moment the process of blending. Let's say that the technician has forgotten to add the active ingredient while loading a blender. A simple study of standard deviations may indicate the time at which all excipients are uniformly blended, when spectral variance reaches a minimum, but not identify the absence of active ingredient or the presence of impurities.

A similar problem could arise if a drug, or an excipient for that matter, were somehow lost in a dead space within the blender or adhered to some surface in the blender through a static charge. On the other hand, a pattern recognition test that recognizes differences in the size, shape, and location of a spectral cluster in a multivariate hyperspace will recognize the absence of active ingredient or any excipient (or errant concentrations).

In the first multiple-blend study, the multiple-sample bootstrap technique correlated with the reference UV assay, correctly identifying that the appropriate potency was not reached until the 20 min time point, although standard deviations leveled off at 10 min. The 10 and 15 min samples had similar percent RSD when compared to the 20 min training group, but showed potencies varying by about 2% from the theoretical concentration, which was reached at 20 min.

In the second part of Wargo's study involving analysis of a single blend over 30 min, the pattern recognition methods and the chi-square method indicated homogeneity is reached at 10 min. The reference results suggested that the blend was acceptable in terms of potency (99.1%) but marginal in terms of uniformity (RSD = 2.4%) at the 10 min time point. By 16 min of blending, both potency (100.0%) and uniformity (RSD = 0.9%) were at optimal levels. Figure 3.3 displays a plot of correlation versus time for the multiple-sample bootstrap method, indicating homogeneity is achieved at 10 min and maintained through the final 30 min sample point. An interesting phenomenon seen in this plot is the drop in correlation at 20 min.

Wargo and Drennen attempted to identify the physical and chemical phenomena responsible for spectral changes after the 20 min time point by examining loadings spectra for each of the first four principal components used in this model. The loadings spectra display the coefficient or weighting given to each wavelength in a particular principal component array. Often, a loadings spectrum will exhibit similarity to the spectrum of an individual component, indicating that the principal component in question will describe changes in the concentration or nature of that component. In this study, the fourth loadings spectrum demonstrated features similar to those of a spectrum from pure magnesium stearate. The potential for decreased powder blend uniformity and related NIR spectral variability following extended blending of formulations containing magnesium stearate appears to be possible in light of the results reported by the following investigators.

More recently, the concept of the net analyte signal (NAS), originally proposed for the determination of multivariate figures of merit [17] and later employed as a preprocessing method for quantitative models [18], was used to characterize blend mixing processes. The NAS represents the part of the signal related to the component of interest that is orthogonal to the subspace formed by the other components of a mixture. A NAS value can be calculated for each sample. Skibsted et al. [19] used the NAS of a "golden batch," considered homogeneous, and used it during subsequent blending experiments to compare the variance of the NAS between the golden batch and the batch to characterize.

Puchert et al. [20] proposed an extension of the use of PCA with the principal component score distance analysis approach. In their study, they calculated the Euclidean distance between two successive spectra considering the first three factors. A moving block standard deviation was then applied onto the distance terms. It is an extension of the work of Storme-Paris et al., who considered each principal component independently [21]. In addition, Puchert et al. used a SIMCA-like approach by developing a successive PCA model with only the most stable spectra in calibration. It allowed them to obtain a better resolution when projecting new samples.

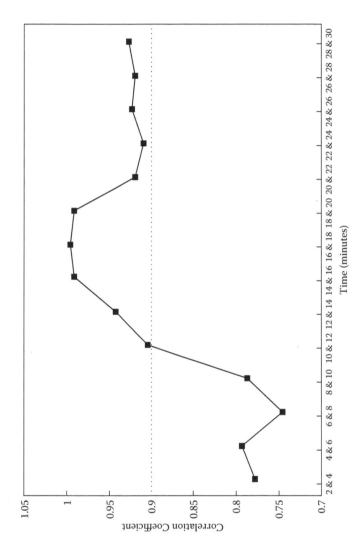

Figure 3.3 Correlation plot for powder blending experiment indicating that content uniformity is reached after 10 min of mixing.

They considered the evolution of the standard deviation and Hotelling's T^2 control charts to determine endpoints.

Near-infrared imaging has also been used to characterize blends. El-Hagrasy et al. [22] used a moving block standard deviation (10 × 10 pixels) at absorption wavelengths related to the molecules of interest to evaluate a blend. The same research group published a later article using a PCA model developed with pure component images to characterize the dispersion of the various components of the mixture [23].

Multivariate Curve Resolution (MCR) was recently used as a qualitative method to monitor blending. In their study, Jaumot et al. [24] used natural constraints in the MCR algorithm (non-negativity for concentration blending profiles and spectra) to resolve the trends of the active ingredient and excipients. The authors suggested the use of pure spectra to improve the fit, which would position MCR at the interface between unsupervised and supervised methods. The performance of MCR was compared in a subsequent study with another qualitative method based on Hotelling's T2 and Q residuals used to determine a distance value from homogeneity [25]. The method, Prototype, compares the evolution of the statistics and determines membership of a particular spectrum to the homogeneous or heterogeneous group based on its distance from a calibration pool constituted of homogeneous spectra.

3.2.2   Quantitative approaches

Shi et al. [26] noted the following: "While it is frequently argued that the concentration variation is implicit in qualitative methods, the expression of variation is explicit in quantitative methods due to the specificity of a calibration model for a particular concentration variation."

In recent years, blend characterization by quantitative means has become particularly popular. Quantitative approaches require a calibration model to capture the concentration variation during the blending process. A quantitative model is able to express blending processes in terms of concentration variation, which is comparable to the standard criteria of current regulatory requirements.

Compared to qualitative methods that aim at detecting changes or rather the lack of change in spectral variation to determine endpoint, the evolution of the predicted values is used when employing quantitative approaches. Partial least-squares (PLS) regression has been the primary method used to develop quantitative models [26–28]. However, other, less popular modeling methods have been employed and shown to perform adequately [29, 30]. Zacour et al. [30] compared PLS to various implementations of classical least-squares regression (involving augmentation procedures of the K-matrix with derived shapes of interferences) and some nonlinear approaches (least-squares support vector machines and artificial neural networks). While results for all methods were very comparable, authors found that classical least-squares-based methods possessed advantages with respect to sensitivity that was lacking for non-linear approaches.

In most situations, reported results focus on the monitoring of the active ingredient alone, as it is often the only parameter that is being measured by wet chemistry. The prediction of the distribution of lubricant (magnesium stearate) was found to be possible at levels as low as 0.5% w/w [31]. Recognizing the importance of excipient homogeneity for the performance of the delivery form, Shi et al. [26] introduced pooled statistics to consider the distribution of not only the active ingredient, but also the excipients.

The root mean square error from the nominal value (RMSNV) is a weighted, cumulative, pooled standard deviation indicator that takes into account the deviation of the predicted concentration of the major components of a mixture to their target concentration, over a given number of rotations. For an n component system, it is defined as

$$RMSNV_{t-i}^{t} = \sqrt{\frac{\sum_{j=t-i}^{t} w_1\left(\hat{Y}_{1j} - Y_{1j}\right)^2 + \sum_{j=t-i}^{t} w_2\left(\hat{Y}_{2j} - Y_{2j}\right)^2 + \dots + \sum_{j=t-i}^{t} w_n\left(\hat{Y}_{nj} - Y_{nj}\right)^2}{i\left(\sum_{j=t-i}^{t} w\right)}}$$

$$(3.2)$$

where w is the weight given to a component, $\hat{Y}_{nj}$ is the predicted value and Y_{nj} is the target value for the nth component, i is the number of data points used in a single window, t is the latest collected data point, and j is an individual prediction time.

The RMSNV allows the evaluation of the trends and distance from target concentrations for not only the active pharmaceutical ingredient, but also the major excipients of the formulation. In the original paper, the endpoint was determined based on the distribution of RMSNV values. In Zacour et al. [30], the root mean square error of calibration for the components of interests was used as the threshold under which the blend was determined homogeneous.

Multivariate Curve Resolution used in the quantitative mode was compared to PLS prediction results [25]. The MCR algorithm used a correlation constraint that related the values from the calculated concentration profiles to the known concentration of the active and other excipients [24].

While also requiring a calibration set, MCR proved it provided comparable results to PLS, while based on Beer's Law and subject to deviations from the Law in the analysis of solids by diffuse reflectance NIR analysis.

3.2.3   Mixing considerations

Shah and Mlodozeniec [32], in their study of surface lubrication phenomena, suggested that during the mixing process, lubricant particles such as magnesium stearate first adsorb onto the surface of individual powder particles or granules, and then, as mixing continues, distribute more uniformly upon the granule surface following delamination or deagglomeration mechanisms. By affecting the surface characteristics of the powder particles, the magnesium stearate may alter the flow properties of the material and affect the apparent bulk volume of the blended material. Moreover, Murthy and Samyn [33] observed that the drug dissolution profiles for lactose–magnesium stearate compacts were related to the degree of shear applied during the mixing process.

Longer periods of shear resulted in extended dissolution times. The authors also noted that powder blend bulk and tapped densities increased as lubricant blend times increased. The increased density was attributed to the improved flow properties of the blend. Although evidence suggests that the spectral variations apparent beyond 20 min in this study are related to the effects of magnesium stearate, further studies are necessary before this can be established with greater certainty.

Krise and Drennen [34] witnessed a similar increase in standard deviations during a blending study that involved only simple binary mixtures of pharmaceutical powders without the inclusion of magnesium stearate. Evidence of this phenomenon in the Krise and Drennen study is likely due to continued shear during prolonged mixing causing a milling effect with corresponding segregation as the particle size distribution changes. Another possible cause is fluctuating electrostatic charge. A cyclic mixing and demixing appears to occur as mixing continues for prolonged periods.

The appearance of different blend profiles for multiple runs of a constant mixture, seen in the Wargo paper, indicates the potential value of monitoring all blending processes. Fluctuating environmental factors such as humidity and temperature, or product factors like inconsistent particle size or particle size distribution, may alter the optimum blend time, even for an established blending process. The use of a rapid and noninvasive NIR method, applied online as described by Sekulic or Shi, requires a minimum of time and permits the ultimate testing by traditional methods. While the production manager waits for wet chemistry results from the laboratory, continued processing of a batch based on the NIR results that were obtained in real time may be allowed.

3.3   Sampling and data handling

A review of the literature reveals literally dozens of approaches to evaluating the analytical data provided in blend studies. Poux and Fayoulle [3] and Puchert et al. [20] have summarized many of the multitudes of statistical mixing indexes used by various authors. Most of the suggested indexes are based on standard deviation. A number of reasons for avoiding such traditional statistics based on standard deviation can be provided, among them the fact that pharmaceutical blends may not contain normal distributions of components. The analyst must remember the assumptions required in analysis of variance, which include normally distributed errors, independence of errors, and equality of variance. Nonnormal data can give rise to incorrect conclusions regarding homogeneity when using traditional statistical techniques based on standard deviation.

Nonparametric statistical algorithms are available for evaluating the nonnormal data arising from blends. In fact, the bootstrap algorithms of Lodder and Hieftje, briefly discussed in the preceding section, were designed to be used in such situations. These bootstrap algorithms make no assumptions regarding the normality of data, taking advantage of the multivariate nature of NIR spectra, and are responsible in part for the success achieved with their use.

Bolton [15] has commented on the comparison of variances for determination of homogeneity. The degree of homogeneity is determined by analyzing the samples from different locations in a mix and calculating the standard deviation. Small standard deviations indicate homogenous blends, whereas larger standard deviations indicate less homogeneity. An F-test or chi-square test of the variance, depending on the sampling protocol, allows determination of homogeneity at the desired confidence level.

Williams [35] has commented on the standard deviation as a measure of quality for a mixture of dry powders. The technique offers certain disadvantages, according to Williams. First, a pharmaceutical mixture is a complex assembly of particles, and it may not be reasonable to assume that it can be adequately described by a single statistic like standard deviation. Sample compositions are often not normally distributed, and although the progress of mixing may be followed by observing changes in standard deviation, the number of samples, or dosage forms, that possess potencies outside of a certain specification cannot be accurately predicted from the standard deviation. A second objection that Williams discussed involved the fact that a standard deviation depends on the size of the sample.

Poux and Fayoulle [3] summarized the results of other researchers investigating the effect of sample size on assessment of the quality of a mixture. The size of the sample must be adapted to the dimensions of the powder material, whose distribution in the mixture must be determined.

The nature of individual particles and their size in comparison with the size of the sample must be considered. The size of a sample should correspond to the use of the mixture. The reason for unit-dose samples in powder blend testing is to ensure that individual doses contain the intended amount of each constituent.

The theory that thieved unit-dose samples will provide the best estimate of homogeneity in a final dosage form may be flawed, however, because of the segregation that occurs as a powder flows into the thief. Identification of the effective sample when analyzing powder blends with NIR will be an important part of developing an optimal methodology and may be possible by one of several approaches [33, 34]. Judiciously selecting sampling optics to provide an illuminated sample of approximately the same size as a dosage form may offer a means of unit-dose sampling without the error associated with thieved samples. Williams's third objection to the standard deviation is that it is misleading when applied to mixtures of different compositions. He suggested, rather, the coefficient of variation, which eliminates the bias from variations in the level of target potency.

An advantage of the NIR method over traditional potency determinations and standard deviation testing is that by providing multivariate data on the physical and chemical nature of a sample, it can generate a much better understanding of the homogeneity, or lack of it, for a given sample. To realize this advantage, the NIR data must be utilized with one of the multivariate pattern recognition algorithms that exploit the multivariate nature of the NIR data. Where traditional methods provide the standard deviation of a single chemical measurement of one constituent, NIR methods can provide over a thousand measurements (absorbance or reflectance values over the range from 1100 to 2500 nm, for example) per sample, giving chemical and physical information about all components of a blend. However, choosing a single NIR wavelength for monitoring the standard deviation of a single absorbance value, in either space or time, negates the advantage possible with multivariate measurements.

Igne et al. [25, 36] studied the relationship between the end-point algorithm employed to determine homogeneity and the window size used to set the sample size included in the decision process. The authors determined that as the window size increased (larger number of spectra included in the decision process), blend homogeneity was harder to reach and required longer blend times. This observation was due to the need for stability in outputs as the number of points taken into consideration increased. The authors further stated that: "The choice of window size should reflect the goal of the blend operation and the method employed to determine homogeneity. A small window size will be particularly sensitive to temporary mixing and demixing phenomena and the scale of scrutiny of the sensors. As the window size increases, more powder is sampled, and the decision is based on a more representative estimate of homogeneity. However, a large window will require a particularly stable blend but provide the best estimate of homogeneity." This statement should however be limited to how variance was handled by the end-point algorithm. Other approaches may affect the window size in a different way.

Regardless of the way in which NIR is used in blend analysis, for observation of a single wavelength or multiple wavelengths, some type of statistic should be applied to the data for decision-making purposes. In doing so, it is wise to consider all sources of variance contributing to the total error of any blend evaluation. Yip and Hersey [37] have indicated that the total variance in a mixture is the sum of variances from mixing, analysis, sampling, and purity:

$$\sigma^2 = \sigma^2_{mixing} + \sigma^2_{analysis} + \sigma^2_{sampling} + \sigma^2_{purity} \tag{3.3}$$

Considering that NIR methods may be used in a noninvasive manner, we can reduce the significant error due to segregation during sampling, leaving us with the following expression for variance using NIR methods:

$$\sigma^2 = \sigma^2_{mixing} + \sigma^2_{analysis} + \sigma^2_{purity} \tag{3.4}$$

Of course, we now assume that some noninvasive sampling protocol can be derived for NIR measurements that give sample spectra that are representative of the blend. A noninvasive sampling method would eliminate or at least reduce the variance from sampling. And if we then assume that our NIR method gives precision that is comparable to some reference technique, which is usually the case, it can be said that the NIR method offers potential for improved estimates of homogeneity over traditional methods. With most substances providing some NIR signature, the NIR method would also be more sensitive to unwanted impurities than, say, a UV or Vis assay.

The choice of an endpoint algorithm must nevertheless be dictated by the product of interest, the particular characteristics the delivery forms should have, and the purpose of blending. No one algorithm is adequate for all formulations. However, with the increase in intricacy of the blending systems (multiple sensors, multiple parameters to follow, etc.), the decision process increases in complexity with not only calibration maintenance issues to consider, but also calibration transfer and the relevance of the desired final product properties [36, 38].

3.4   Segregation, demixing, and particle size

The particle properties that cause segregation, mentioned earlier, tend to prevent effective mixing within a blender and also cause segregation of well-mixed blends. Unfortunately, segregation of blended powders or granulation is likely to occur during the emptying of a mixer, during the subsequent batch transfer or handling, or even within a capsule-filling machine or tablet press. This segregation occurs due to several mechanisms, including trajectory segregation, percolation, and the rise of coarse particles by vibration [35]. Near-infrared methods can detect segregation of powders with the same tests used to identify homogeneity. Again, one of the main advantages is that NIR techniques can identify segregation problems through recognition of chemical and physical changes in a blend. For example, let's say that during the transfer of a blended batch to the compression area, the inherent vibration has caused the smaller particles or fines to settle and the larger particles to rise to the top of a blend. Near-infrared spectra should show both a change in the chemical composition and a change in the particle size. Changing chemical composition that will be identified with NIR methods of blend monitoring could be easily applied to test a batch before initiation of tablet compression, for example, to rule out segregation that might have occurred during blender discharge or batch transfer.

Although the detection limits for NIR methods are much higher than those of many other common methods, typically in the neighborhood of 0.1% due to the low absorptivity of most compounds in the NIR region of the electromagnetic spectrum, NIR should prove to be more sensitive to segregation than expected. Free-flowing powders, segregating due to particle size or particle density differences, would provide both chemical and physical means for the NIR spectrometer to recognize a loss of homogeneity. We assume that any sample from a well-mixed powder blend is going to provide a similar distribution of the various components of the blend.

Conversely, samples arising from blends that have segregated to any significant extent will exhibit unique distributions of active ingredients and excipients. Think for a moment of a powder blend that will segregate a free-flowing powder blend with particles of different sizes. Using the NIR method, not only can the sample constituent concentrations be predicted as different from those desired in a homogenous blend, but the shifting particles can be identified as well because of the ability to study surface phenomena. The NIR method thereby bolsters the potency measurement with the possibility of detecting the baseline shifting that occurs with changes in particle size.

A NIR sensor on the feed hopper of a tablet press or capsule-filling machine might be used to ensure that segregation does not occur as powder or granulation is fed through the equipment. Just one example of a situation where such a sensor may be valuable is presented by Johansen et al. [39] in their discussion of the segregation and continued mixing of powders in an automatic capsule-filling machine. They recognized problems of segregation and overmixing of a powder mixture as the causes for dose variations and inconsistent dissolution rate. The investigators modified the capsule-filling machine to achieve the desirable dosage form, in regard to dose and performance. Johansen et al. replaced the hopper with one of a new design and removed the mixing blade to improve powder flow and achieve satisfactory single-dose variation and dissolution rate.

3.5 Conclusion

The ultimate methodology for successful NIR monitoring of mixing at production scale has yet to be identified. To get the most from this technology, which is relatively new to the pharmaceutical industry, will require taking advantage of the multivariate nature of the technique. However, considering that pharmaceutical manufacturers have for years been successfully producing blends that meet United States Pharmacopeia requirements while using standard statistical techniques that rely on single parameters such as standard deviation, it may not be necessary to use the more complicated algorithms for multivariate data analysis. The noninvasive nature of NIR offers an advantage even when using single-wavelength measurements with traditional statistical evaluation of the data.

References

1. J. K. Drennen and R. A. Lodder, NIR Analysis of Ointments, presented at 2nd Annual NIRSystems Symposium on Pharmaceutical Applications of Near-IR Spectrometry, College of St. Elizabeth, Convent Station, NJ, October 25–26, 1989.
2. J. K. Drennen and R. A. Lodder, Pharmaceutical Applications of Near-Infrared Spectrometry, in *Advances in Near-IR Measurements*, ed. G. Patonay, JAI Press, Greenwich, CT, 1993.
3. P. Poux and J. Fayolle, Bertrand, Bridoux, Bousquet, Powder Mixing: Some Practical Rules Applied to Agitated Systems, *Powder Technol.*, 68, 213 (1991).
4. J. B. Gray, [Textbook] *Chem. Eng. Prog.*, 53, 25 (1957).
5. M. D. Ashton, C. Schofield, and F. H. H. Valentin, The Use of a Light Probe for Assessing the Homogeneity of Powder Mixtures, *Chem. Eng. Sci.*, 21, 843 (1966).
6. E. W. Ciurczak, Use of Near Infrared in Pharmaceutical Analyses, *Appl. Spectrosc. Rev.*, 23, 147 (1987).
7. F. C. Sanchez, J. Toft, B. Van den Bogaert, D. L. Massart, S. S. Dive, and P. Hailey. Monitoring Powder Blending by NIR Spectroscopy, *Fresenius J. Anal. Chem.*, 352, 771 (1995).
8. W. Windig and J. Guilment, Interactive Self-Modeling Mixture Analysis, *Anal. Chem.*, 63, 1425 (1991).

9. S. S. Sekulic, H. W. Ward II, D. R. Brannegan, E. D. Stanley, C. L. Evans, S. T. Sciavolino, P. A. Hailey, and P. K. Aldridge, On-Line Monitoring of Powder Blend Homogeneity by Near-Infrared Spectroscopy, *Anal. Chem.*, 68, 509 (1996).

10. S. S. Sekulic, J. Wakeman, P. Doherty, and P. A. Hailey, Automated System for the Online Monitoring of Powder Blending Processes Using Near Infrared Spectroscopy. Part II: Qualitative Approaches to Blend Evaluation, *J. Pharm. Biomed. Anal.*, 17, 1285 (1998).

11. R. Maesschalck, F. C. Sanchez, D. L. Massart, P. Doherty, and P. Hailey, On-Line Monitoring of Powder Blending with Near-Infrared Spectroscopy, *Appl. Spectrosc.*, 52, 725 (1998).

12. D. A. Wargo and J. K. Drennen, Near-Infrared Spectroscopic Characterization of Pharmaceutical Powder Blends, *J. Pharm. Biomed. Anal.*, 14, 1415 (1996).

13. R. A. Lodder and G. M. Hieftje, Quantile BEAST Attacks the False-Sample Problem in Near-Infrared Reflectance Analysis, *Appl. Spectrosc.*, 42, 1351 (1988).

14. R. A. Lodder and G. M. Hieftje, Detection of Subpopulations in Near-Infrared Reflectance Analysis, *Appl. Spectrosc.*, 42, 1500 (1988).

15. S. Bolton, Examination of Raw Materials with NIRS, in *Pharmaceutical Statistics*, 2nd ed., Marcel Dekker, New York, 1990.

16. B. Efron and R. Tibshirani, Statistical Data Analysis in the Computer Age, *Science*, 253, 390 (1991).

17. A. Lorber, Error Propagation and Figures of Merit for Quantification by Solving Matrix Squations, *Anal. Chem.*, 58, 1167 (1986).

18. H. C. Goicoechea and A. C. Olivieri, Chemometric Assisted Simultaneous Spectrophotometric Determination of Four-Component Nasal Solutions with a Reduced Number of Calibration Samples, *Anal. Chim. Acta*, 453, 289 (2002).

19. E. T. S. Skibsted, H. F. M. Boelens, J. A. Westerhuis, D. T. Witte, and A. K. Smilde, Simple Assessment of Homogeneity in Pharmaceutical Mixing Processes Using a Near-Infrared Reflectance Probe and Control Charts, *J. Pharm. Biomed. Anal.*, 41, 26 (2006).

20. T. Puchert, C. V. Holzhauer, J. C. Menezes, D. Lochmann, and G. Reich, A New PAT/QbD Approach for the Determination of Blend Homogeneity: Combination of On-Line NIRS Analysis with PC Scores Distance Analysis (PC-SDA), *Eur. J. Pharm. Biopharm.*, 78, 173 (2011).

21. I. Storme-Paris, I. Clarot, S. Esposito, J. C. Chaumeil, A. Nicolas, F. Brion, A. Rieutord, and P. Chaminade, Near InfraRed Spectroscopy Homogeneity Evaluation of Complex Powder Blends in a Small-Scale Pharmaceutical Preformulation Process, a Real-Life Application, *Eur. J. Pharm. Biopharm.*, 72, 189 (2009).

22. A. S. El-Hagrasy, H. R. Morris, F. D'Amico, R. A. Lodder, and J. K. Drennen 3rd, Near-Infrared Spectroscopy and Imaging for the Monitoring of Powder Blend Homogeneity, *J. Pharm. Sci.*, 90, 1298 (2001).

23. H. Ma and C. A. Anderson, Characterization of Pharmaceutical Powder Blends by NIR Chemical Imaging, *J. Pharm. Sci.*, 97, 3305 (2008).

24. J. Jaumot, B. Igne, C.A. Anderson, J.K. Drennen, and A. De Juan. Blending Process Modeling and Control by Multivariate Curve Resolution, *Talanta*, 117, 492–504 (2013).

25. B. Igne, A. De Juan, J. Jaumot, J. Lallemand, S. Preys, J.K. Drennen, and C.A. Anderson. Modeling Strategies for Pharmaceutical Blend Monitoring and Endpoint Determination by Near-Infrared Spectroscopy, *Intl. J. Pharm.*, 473, 219–231 (2014).
26. Z. Shi, R. P. Cogdill, S. M. Short, and C. A. Anderson, Process Characterization of Powder Blending by Near-Infrared Spectroscopy: Blend End-Points and Beyond, *J. Pharm. Biomed. Anal.*, 47, 738 (2008).
27. O. Berntsson, L. G. Danielsson, B. Lagerholm, and S. Folestad, Quantitative In-Line Monitoring of Powder Blending by Near Infrared Reflection Spectroscopy, *Powder Technol.*, 123, 185 (2002).
28. M. Popo, S. Romero-Torres, C. Conde, and R. J. Romañach, Following the Progress of a Pharmaceutical Mixing Study via Near-Infrared Spectroscopy: Blend Uniformity Analysis Using Stream Sampling and Near Infrared Spectroscopy, *AAPS PharmSciTech*, 3, 1 (2002).
29. H. Wu, M. Tawakkul, M. White, and M. A. Khan, Quality-by-design (QbD): An Integrated Multivariate Approach for the Component Quantification in Powder Blends, *Int. J. Pharm.*, 372, 39 (2009).
30. B. M. Zacour, B. Igne, J. K. Drennen III, and C. A. Anderson, Efficient Near-Infrared Spectroscopic Calibration Methods for Pharmaceutical Blend Monitoring, *J. Pharm. Innov.*, 6, 10 (2011).
31. N. H. Duong, P. Arratia, F. Muzzio, A. Lange, J. Timmermans, and S. Reynolds, A Homogeneity Study Using NIR Spectroscopy: Tracking Magnesium Stearate in Bohle Bin-Blender, *Drug Dev. Ind. Pharm.*, 29, 679 (2003).
32. A. C. Shah and A. R. Mlodozeniec, Mechanism of Surface Lubrication: Influence of Duration of Lubricant-Excipient Mixing on Processing Characteristics of Powders and Properties of Compressed Tablets, *J. Pharm. Sci.*, 66, 1377 (1977).
33. K. S. Murthy and J. C. Samyn, Effect of Shear Mixing on In Vitro Drug Release of Capsule Formulation Containing Lubricants, *J. Pharm. Sci.*, 66, 1215 (1977).
34. J. P. Krise and J. K. Drennen, presented at American Association of Pharmaceutical Scientists Annual Meeting and Exposition, San Antonio, TX, November 1992, paper APQ 1108.
35. J. C. Williams, The Mixing of Dry Powders, *Powder Technol.*, 2, 13 (1968–1969).
36. B. Igne, S. Talwar, J.K. Drennen, and C.A. Anderson. On-line Monitoring of Pharmaceutical Materials Using Multiple NIR Sensors—Part II: Blend End Point Determination, *J. Pharm. Innov.*, 8, 1, 45–55 (2013).
37. C. W. Yip and J. A. Hersey, Perfect Powder Mixtures, *Powder Technol.*, 16, 189 (1977).
38. B. Igne, B.M. Zacour, Z. Shi, S. Talwar, J.K. Drennen, and C.A. Anderson. On-line Monitoring of Pharmaceutical Materials Using Multiple NIR Sensors—Part I: Blend homogeneity, *Journal of Pharmaceutical Innovation*, 6(1):47-59 (2011).
39. H. Johansen, I. S. Anderson, and H. Leedgaard, Identification of Actives in Multi-Component Pharmaceutical Dosage Forms via Near-IR Reflectance Analysis: Segregation and Continued Mixing in an Automatic Capsule Filling Machine, *Drug Dev. Ind. Pharm.*, 15, 477 (1989).

chapter 4

Granulation, drying, and coating

4.1 Introduction

Near-infrared (NIR) spectroscopy has proven valuable for monitoring intermediate pharmaceutical products and processes as well as finished products. The technique offers significant advantages over traditional destructive analytical methods for the monitoring of pharmaceutical granulation processes, as it does for other dosage forms and unit operations. This chapter discusses NIR spectroscopy as applied to traditional granulation and to the related processes of drug layering and film coating of beads or granules. The parameters of interest include potency, moisture content, and coating level. Examples to be discussed in this chapter include (1) the monitoring of granulation and drying, (2) the layering of drug suspension, (3) polymer film coating, and (4) tablet coating. In each case, data exist to suggest that NIR methods can be applied at-line or online to improve the ability for monitoring and controlling such processes. The ability to rapidly determine the endpoint of such processes can simplify process development, scale-up, and production.

The production of solid dosage forms such as tablets or capsules often requires a granulation step. Granulation serves several purposes, which include provision of improved flow characteristics in comparison to finely divided powders, improved compaction tendencies, and achieving and maintaining content uniformity. When this process is carried out in the traditional manner of wet granulation, the dry solid products are typically blended before addition of an aqueous solution containing binder. The binder is added during continued mixing in one of a variety of possible mixers and serves to bind the solid particles into a mass in a process called wet massing. A screening step usually follows, with formation of granules of a particular and consistent size. The granulation must eventually be dried. Drying is achieved by one of several possible methods, including tray drying, fluid bed drying, or even microwave drying. It is during this drying process that NIR has proven its value for control of the granulation process due to its ability to monitor moisture content online. The residual moisture content is known to affect the ultimate performance of a tablet dosage form [1], making control of the drying process of critical importance.

Of further importance, Chaudry and King have witnessed the migration of active ingredient in the granulation step as being responsible for nonuniform tablets [2]. Sodium warfarin tablets prepared by a wet granulation method exhibited unsuitable content uniformity due to the migration of active ingredient during drying. The ability to monitor content uniformity by NIR may allow such migration of drug to be recognized when the drying process is monitored directly.

With the possibility of monitoring content uniformity and particle size as well as moisture content, NIR should also be valuable for control of the wet massing step during granulation. Although granulation is not usually considered to be a mixing procedure, content uniformity is certainly a concern, and NIR may be used to monitor the uniformity of granulations with a variety of conformity tests. The conformity index of Plugge and Van der Vlies, discussed in this chapter, is another possible approach for NIR testing of granulations. Such phenomena indicate the potential value of online control mechanisms like NIR. The methods discussed in Chapter 3 on mixing should also be applicable.

Production of sustained or controlled release capsule dosage forms will often involve pharmaceutical pelletization technology. Pelletization involves the layering of a drug onto inert beads, followed by coating various fractions of the batch with different levels of polymeric coating material. By combining fractions of pellets with variable levels of a polymer coating, the profile of drug release from the final dosage form can be controlled very closely. The processes of layering drug suspension and polymer film coating can be monitored and controlled by NIR methods, to provide accurate prediction of potency and applied polymer solids during rotogranulation and fluid bed coating operations, thus making it possible to closely monitor pellet processing and achieve reproducible endpoint product characteristics. The same argument can be made for the monitoring and control of the tablet-coating process.

4.2   Monitoring granulation and drying

Abundant evidence exists to display the value of NIR for moisture determinations. Water has a characteristic spectrum, as seen in Figure 4.1. The presence of moisture will tend to alter the spectrum of samples in many cases, hiding weak absorbance bands and shifting the location of other bands. The spectral variations that occur with changing moisture content are due to the significant absorbance of water across the NIR region and also because of the hydrogen bonding between water and many common functional groups of pharmaceutical materials.

Water is also a common source of error in NIR measurements. Variable water content can cause significant changes in sample spectra, and unless such moisture variations are included in a calibration, NIR-predicted

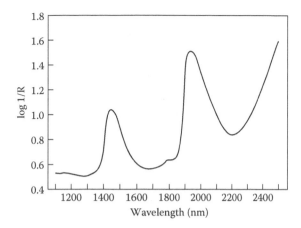

Figure 4.1 NIR spectrum of water.

values for potency, coating level, and other parameters may be in error. While the goal of NIR measurements in monitoring traditional granulation processes is likely to be moisture determination, when other parameters are under study, it would be wise to provide humidity control. For example, during the coating of drug-loaded nonpareil seeds with polymer in a fluid bed system, it would be wise to control the humidity of incoming air. Failure to do so could lead to significant error in the NIR-predicted coating weights, not to mention poor control of the coating process.

As an example of the error that may be induced in NIR measurements by failure to control humidity, fluctuations in the concentration of atmospheric water vapor generated by an air conditioner were identified as a source of noise in NIR spectrometers in a 1987 paper [3].

Ellis and Bath studied changes in the NIR absorption spectra of water bound to gelatin in a 1938 paper [4]. The intensity of the first overtone N–H band at 1500 nm was found to be decreased by the addition of water. The combination N–H bands at 2050 and 2180 nm were also partially attenuated. The 1720 and 2280 nm C–H bands did not change in intensity, but the 1720 nm band shifted to a higher frequency upon binding with water. This was due to the C–H groups occurring in a more vapor-like state with water molecules attached only to the polar groups of the gelatin molecules, according to the authors. The authors also saw loss of some absorption on each side of the 1440 and 1930 nm water bands upon binding with water, reflecting a reduction in the number of unperturbed vapor-like molecules and in the number of highly perturbed molecules involving three and four hydrogen bonds.

In 1968, Sinsheimer and Keuhnelian used the 1900 nm absorption band of water to quantify water with the same accuracy as Karl Fischer titration [5]. This is a combination band of fundamental stretching and bending vibrations of water [6]. Water was observed in several systems, including a hydrophilic petrolatum emulsion, a micelle system, and in organic solvents. Water of hydration of ammonium oxalate and sodium tartrate was determined after extraction with methanol. Solvents that are strongly hydrogen bonding work best for moisture determination due to the greater miscibility with water. The dehydration of cholesterol was followed by observing the change in absorbance at 1920 nm after dissolving samples in pyridine.

In 1970, Warren et al. addressed the determination of trace amounts of water in mono-, di-, and triglycerides by NIR spectrometry [7]. Using chloroform as a solvent, the investigators studied the combination band of water occurring at 1896 nm in the presence of the −OH groups on glycerides to determine water down to 0.05%.

Whitfield used NIR to determine the concentrations of moisture and lincomycin in animal feed in a 1986 publication [8]. This was the first NIR method accepted by the Food and Drug Administration. The method was part of a veterinary product application. Using a loss on drying (LOD) method to test the moisture concentration of calibration samples, a two-wavelength step-up regression was developed to correlate the LOD measurements with NIR spectra. The wavelength selection algorithm chose 1640 and 1896 nm for calibration. While the 1640 nm measurement is a baseline determination, the 1896 nm measurement is obviously providing information regarding the 1930 nm water band and is likely shifted from the expected 1930 nm due to matrix effects.

Kamat et al. first reported the NIR determination of residual moisture in intact lyophilized products [9]. Kamat and coworkers made measurements through the intact glass vials using a diffuse reflectance fiber optic probe. By scanning through the bottom of lyophilization vials, they studied moisture concentrations over the range of 0.72 to 4.74%. Water peaks were observed in the 1450 and 1930 nm regions. With Karl Fischer as the reference method, the standard error of prediction for this study was 0.27%. The study showed that NIR is more sensitive to water than Karl Fischer when water peaks were observed after the wet method displayed 0% water concentration in the samples. Others have reported similar good results for the noninvasive determination of moisture in lyophilized pharmaceuticals [10, 11].

Plugge and van der Vlies have discussed the conformity index (CI) for NIR analysis of ampicillin trihydrate [12, 13]. The CI is a metric used to determine the degree of conformity of a sample or batch with standards of known and acceptable quality. To use the CI, reference spectra are first collected and baseline-corrected using a second-derivative or multiplicative scatter correction (MSC) spectrum. At every wavelength across the spectrum, the average absorbance and standard deviation are calculated for the baseline-corrected reference spectra, resulting in an average spectrum and a standard deviation spectrum.

The spectrum of a sample to be tested is recorded and baseline-corrected by the appropriate method. At each wavelength, a Q_λ value is calculated according to the following equation, where the absolute difference between the absorbance of the sample spectrum (A) at some wavelength (λ) and the average reference spectrum ($\bar{A}$) is divided by the standard deviation of the reference spectra at the same wavelength (σ):

$$Q_\lambda = \frac{|A - \bar{A}|}{\sigma} \tag{4.1}$$

The CI is the maximum Q_λ for that sample. The CI values can be plotted in traditional control charts to monitor a process. The CI described by Plugge and van Der Vlies is essentially the wavelength distance pattern recognition method available commercially [14].

Gemperline et al. discussed an improved methodology for wavelength distance measurements in an article comparing Mahalanobis distance and SIMCA [15]. Gemperline et al. have normalized the CI value to account for sample size and to allow for qualitative decisions to be made according to a probability threshold by comparison to critical t values, rather than a Q_λ threshold.

On large data sets (greater than 14 samples) the SIMCA and Mahalanobis distance methods perform better than the wavelength distance method because they use estimates of the underlying population variance/covariance matrix for sample classification. As the size of the training set decreases, the accuracy of the population variance/covariance matrix estimates decreases, and performance of the Mahalanobis distance and SIMCA methods is reduced. In such cases where small training sets are used, the wavelength distance method may perform better because its univariate means and standard deviations are likely estimated with more certainty than the multivariate means and variance/covariance matrices used by the Mahalanobis distance and SIMCA methods.

For analysis of more complicated tablet matrices, it is unlikely that training sets of 14 or fewer samples would suffice in the development of robust models for quantitation or qualitative discrimination. Considering a typical tablet with, say, four ingredients, one might determine an approximate number of training samples needed to develop a good model for quantitation or discriminant analysis. Assuming one needs about 10 training samples to describe each variable and 10 samples for a calibration constant, at least 70 training samples would be necessary if hardness and moisture are included as variables. As a result, it is expected that the performance of multivariate pattern recognition methods such as Mahalanobis distance, SIMCA, or bootstrap error-adjusted single-sample technique (BEAST) should surpass that of the wavelength distance pattern recognition method for qualitative tablet analysis. Gemperline et al.'s results suggest that the performance of the wavelength distance pattern recognition method (and the CI method) offered poor results in detection of borderline samples and samples adulterated with low levels of contamination.

Although the conformity index should provide a measurement of the total quality of the product, including the adherence to moisture specifications, Plugge and van der Vlies also developed a quantitative calibration for water. With Karl Fischer as reference, their calibration covered the range from 7.1 to 11.6% moisture and used 1642 and 1930 nm to give suitable linearity.

Qualitative tests for product conformity to specified limits on potency, moisture, and other parameters may offer several advantages over the more traditional quantitative calibration approaches. Fortunately, today's pharmaceutical production environment provides products that meet exacting standards. The actual variability in potency for a tablet dosage form, as an example, is minimal. A consequence of such tight tolerances, however, is that it is sometimes impossible to develop a meaningful quantitative calibration with traditional regression techniques using production samples because the range of potencies is so narrow.

The analyst may then attempt to use laboratory-prepared tablets to develop a calibration over a wider range of potencies. This is generally not an acceptable solution, however, because of the difference in samples prepared on lab-scale versus production-scale equipment. A qualitative test, on the other hand, whether in the form of a pattern recognition algorithm or a conformity test, offers an advantage in that the range of potencies, moisture values, or all parameters of interest for that matter, are unimportant in developing a model to predict good or bad samples, as long as all potential sources of variation are included in the model.

Caution must be used in developing quantitative calibration models for NIR spectroscopy. The analyst must not forget the basics, including statistical assumptions, of regression modeling. Assumptions include (1) for each specific X there is a normal distribution of Y from which sample values of Y are assumed drawn at random, (2) the normal distribution of Y corresponding to a specific X has a mean that lies on the population regression line, and (3) the normal distributions of Y for particular X are independent with constant variance. Certainly, also, a linear relationship must exist between Y and X. The common transformation of NIR reflectance data to log $1/R$ as an approximation of absorbance generally gives an approximately linear relationship over a useful range of constituent concentrations. However, when developing calibrations, the analyst must be constantly alert to the absence of linear correlation between Y and X. This is of most concern when the attempted calibration is over large concentration ranges (>10%). This is a situation that is very likely to occur when monitoring granulations during drying where a very broad range of moisture concentration is possible and can lead to poor prediction results.

On-line moisture detection for a microwave vacuum dryer was the topic of a publication by White [16]. The paper described a method for determining the moisture endpoint in a microwave dryer using NIR spectroscopy. White's calibration equation used NIR absorbance measured at 1410, 1630, and 1930 nm. The NIR-predicted results were within 1% of the Karl Fischer reference values for samples with less than 6% moisture, providing a standard error of prediction (SEP) of 0.6%. For moisture levels above 6%, a bias existed in the NIR data, possibly due to sampling limitations, the authors suggested. Because measurements could only be taken at the edge of the dryer in the online system, the measurements probably did not accurately represent the moisture content of the bulk of the material at the early time points in the process. White also suggested that because of the broad range of moisture contents, from 0.7 to 25.9%, a second calibration equation may reduce the error of the NIR method. A second calibration would reduce the error occurring from the nonlinearity often seen over such a broad range of constituent concentrations. The calibration equation was not sensitive to variations in the sample other than water content. Changes in drug content, for example, did not affect the prediction of moisture content.

White witnessed distinctive patterns in the drying curves observed with the NIR method. Drying occurred rapidly at the onset because the moisture level dropped rapidly at the surface of the material where the NIR measurements were taken. When the impeller stirred the material, fresh granulation with a higher moisture content was moved to the window surface, causing a jump in the NIR-predicted moisture content. As the bulk of the material began to dry, the variation due to stirring decreased until, near the end of the drying process, the drying curve was smooth. The smooth drying curve indicated homogeneity in terms of moisture content. Sanghvi et al. have also used NIR with qualitative pattern recognition methods to monitor moisture homogeneity in their studies of bioadhesive polymers [17].

Since these early experiments, a significant amount of work has been done to monitor granulation and drying by NIR. At the 1996 annual meeting of the American Association of Pharmaceutical Scientists, Han and Oh presented their work regarding online moisture measurement during granulation in a fluid bed system [18]. Their work involved the use of an externally mounted single-wavelength NIR instrument to monitor drying directly through the glass window of a 300 kg fluid bed granulator. Close control of moisture was critical for this tablet formulation because high moisture content led to accelerated degradation of the active ingredient. Han and Oh found that the loss on drying and Karl Fischer methods could be easily replaced by the NIR method. The NIR method offered an advantage over the other methods in that it provided a continuous, real-time measurement providing greater batch-to-batch consistency. By avoiding the withdrawal of samples with the NIR method, required for the other methods, the remoisturizing/redrying steps were eliminated. Overall, a significant savings in time and expense was realized while providing more precise control over the granulation process.

Frake et al. [19] used the measurement technique online to monitor a fluid bed granulation process, adjust it to the unique characteristics of every batch, and determine the endpoint. A similar study by Rantanen et al. [20] showed, in addition to the suitability of NIR to monitor fluid bed granulation online, the possibility to use the technique to monitor drying operations. Authors confirmed the agreement between NIR predictions and Karl Fischer titration. The same research group confirmed earlier findings by Ciurczak et al. [21] about the possibility to predict granule particle size by NIR during granulation [22]. Similar results were obtained by Nieuwmeyer et al. [23], who were able to develop a prediction model using partial least-squares for the prediction of median granule size. The prediction of bulk density was also found possible [24]. Chablani et al. [25] discussed the implementation of NIR for moisture analysis in a continuous granulation-drying milling process.

Zhou et al. [26] discussed the capacity of using principal component analysis to differentiate between the removal of surface water and bond water, and the possibility to determine precise endpoints in situations where several hydrates coexist. Note that the ability of NIR spectroscopy to determine the presence of polymorphs and pseudopolymorphs will be discussed in Chapter 5.

Li et al. [27] used NIR spectroscopy to troubleshoot and optimize a wet granulation process. During development phases, authors realized that out-of-trend content uniformity values were obtained for some sieve cuts due to the interaction of the excipients and the binding solution during the coalescence process. A qualitative method was put in place to determine within different sieve fractions the change in content uniformity. Other process optimization methods involving NIR were described by Miwa et al. [28–30]. They used moisture levels predicted by NIR for independent components to estimate a lower and an upper limit for the amount of water needed to granulate a multicomponent system.

Considerations of model validation for the prediction of water loss during drying were studied by Peinado et al. [31]. Authors describe the approach they took to determine the figures of merit required by external guidelines and discuss some of the interpretations of the published methods to accommodate the accuracy of the endpoint determination.

4.3 Coating and pelletization

Near-infrared spectroscopy has been used to monitor coating process and quality of tablets and pellets. Andersson et al. [32] collected the off-line spectra of tablets coated during various amounts of time and predicted the thickness of coating with a limit of quantification around 0.2 mm. Intertablet coating variation was studied off-line by Moes et al. [33]. In both studies, NIR was found to be a suitable method to monitor the coating process and was a tool of choice to determine if process variations existed between manufacturing scales. Gendre et al. developed an extensive study in which the mass of coating material and film thickness was monitored in-line by principal component analysis and partial least-squares [34]. Authors found that NIR results were comparable to terahertz pulse imaging measurements. Möltgen et al. compared the usefulness of hyperspectral imaging along with off-line and in-line NIR measurements to control a coating process [35]. The images allowed a better understanding of the coating variability within each tablet.

The coating in pan and fluid bed is a chaotic process. Tablets are in constant movement, and significant sampling problems exist due to the difficulty to ensure the correct positioning of samples in front of the probe during online measurements. To bypass this issue, Lee et al. [36] describe a method based on averaging and clustering to isolate the most meaningful spectra. The proposed method was able to yield coating thicknesses as good as 3% deviated from actual measurements.

Kato et al. [37] investigated the influence of granule shapes and the ability of NIR spectroscopy to monitor their coating process. As coating proceeds, core features become less distinguishable. For spherical particles, they found that it was possible to monitor the coating process, while for cylindrical granules, the coating layer became too thick and the coating media based on titanium dioxide scattered all the light and did not allow the characterization of the granule coating process.

Pelletization technology has provided pharmaceutical formulation scientists with tremendous flexibility during solid oral dosage form development. Unfortunately, pharmaceutical pelletization may often involve lengthy and expensive manufacturing processes. The development of analytical methods to evaluate pellet characteristics at-line or online during the manufacturing process may reduce production cycle downtime associated with the acquisition of laboratory test results and allow product uniformity to be assessed prior to completing the manufacture of an entire batch.

Wargo and Drennen used NIR spectroscopy to monitor the potency and applied polymer solids content of controlled release beads at-line during laboratory- and pilot-scale rotogranulation and coating operations [38]. An aqueous suspension of diltiazem HCl, PVP K29/32, and micronized talc was layered onto nonpareil seeds. The drug-layered beads were subsequently coated with Eudragit RS30D using a Wurster column. A Glatt GPCG-3 and a Vector FL60 were utilized in laboratory and pilot studies, respectively. Unit-dose samples were collected at various time points during each operation and scanned at-line using a grating-type NIR spectrometer. Calibrations for potency and applied polymer solids were developed using principal component analysis (PCA), partial least-squares (PLS), and single-wavelength regression models. Using the established calibrations, it was possible to accurately predict potency and applied polymer solids at-line during rotogranulation and coating operations, thus making it possible to closely monitor pellet processing and achieve reproducible endpoint product characteristics.

4.3.1 Rotogranulation studies

At the laboratory scale, triplicate 3 kg batches of beads containing 15, 30, and 55% active drug were prepared by suspension layering a 40% (w/w) aqueous suspension consisting of diltiazem HCl (88%), PVP K29/32 (6%), and micronized talc (6%) onto 25–30 mesh nonpareil seeds. Pilot-scale studies were conducted using the 55% active drug formulation. Five 30 kg batches were processed using the same formula utilized in the laboratory-scale experiment. During the drug layering process, unit-dose samples (500–1000 mg) were collected from 96 to 106% theoretical potency using a sample thief. The samples were screened through #16 and #30 mesh screens and transferred to glass sample vials prior to near-infrared analysis. Diltiazem HCl content was determined by high-performance liquid chromatography (HPLC) according to the United States Pharmacopeia (USP XXIII) official monograph for diltiazem HCl extended release beads.

The 55% diltiazem HCl pellets produced in the laboratory-scale rotogranulation studies were subsequently coated with a 20% w/w aqueous dispersion of Eudragit RS30D, triethylcitrate (20% of polymer solids), and silicon dioxide. All coating studies utilized the GPCG-3 fitted with a 6 in. Wurster column insert. During the coating process, unit-dose samples (500–1000 mg) were collected from 10 to 20% theoretical applied solids using a sample thief. The samples were screened through #16 and #30 mesh screens and transferred to glass sample vials prior to near-infrared analysis.

Near-infrared reflectance spectra for each sample were collected in triplicate by scanning through the base of the sample vials. Individual sample vials were inverted and rotated 120° between scans to ensure representative spectra. The triplicate scans were then averaged to obtain one spectrum for each bead sample. A variety of data preprocessing and calibration routines were evaluated in the development of quantitative regression models for determination of pellet potency and applied

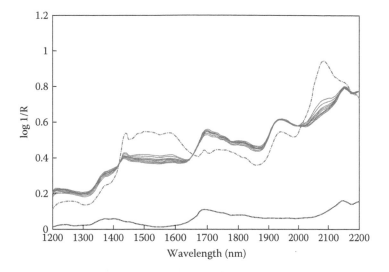

Figure 4.2 Multiplicative scatter corrected NIR spectra of suspension-layered pellets and individual ingredients: diltiazem HCl (dash-dot trace), nonpareil seeds (dot trace), and suspension-layered pellets containing 80–100% of theoretical potency (solid traces).

polymer solids. A qualitative assessment of coated beads was also performed using a bootstrap pattern recognition algorithm.

Figure 4.2 displays MSC-corrected spectra of individual ingredients and suspension-layered pellets. Pellet spectra correspond to samples collected from 80 to 100% theoretical potency based on applied suspension solids. As the layering process proceeds, absorbance decreases in spectral regions characteristic of the nonpareil seeds (1400 to 1600 nm and 2000 to 2400 nm) and increases due to the influence of the applied drug in the 1700 to 1900 nm region.

Independent calibrations for potency were developed for each strength of rotogranulated beads prepared in the laboratory. Thirty-six samples, collected from the first two batches processed for each strength category, were used in the calibration development. Eighteen samples from a third processing run were used to test the calibration. The calibration results for 15, 30, and 55% diltiazem beads demonstrate that near-infrared spectroscopy can be a useful analytical tool for predicting pellet potency over a wide range of drug concentrations. Using principal component regression (PCR), partial least-squares, and single-wavelength calibration models, the standard error of prediction (SEP) values ranged from 0.59 to 1.08%.

Because of the favorable results obtained in the laboratory-scale study, an experiment was conducted to assess the ability to predict the potency of pilot-scale rotogranulated beads using data from the experiments

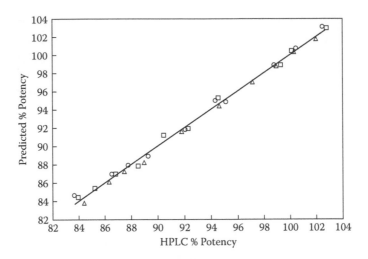

Figure 4.3 Scatterplot indicating the ability to predict pilot-scale potency using laboratory-scale data. $n = 36$, $r^2 = 0.98$, SEP = 1.41%.

performed in the laboratory. The calibration developed for the 55% laboratory-scale beads was used to predict the potency of a 30 kg pilot batch. Figure 4.3 shows a plot of predicted versus actual potency. Prediction error for this study, although acceptable, was slightly higher than in the laboratory study. This error may be attributed to differences in surface characteristics and density between laboratory- and pilot-scale beads.

The application of NIR for predicting the endpoint of clinical or production-scale batches based on the NIR calibration models developed in the laboratory may prove to be very important to the pharmaceutical manufacturer. When drug loading or polymer coating efficiencies vary, as they do, from lab-scale equipment to the larger equipment used for manufacturing, the availability of a rapid online or at-line measurement for prediction of a process endpoint may save many millions of dollars in time and materials. Although the standard error of prediction may not match the values obtained when calibration and validation data arise from product manufactured on the same equipment, the results may be sufficient, and fit for purpose, to prevent the loss of initial batches of clinical or production lots of product and also save valuable time.

Studies were also conducted to assess the ability to monitor clinical batch pellet production and predict endpoint pellet potency. Two 30 kg pilot-scale batches were processed in order to optimize scale-up process parameters. From each batch, 18 samples ranging from 90 to 100% theoretical potency were analyzed and used to develop potency calibrations. The regression results for PCR, PLS, and single-wavelength calibration provided SEP values ranging from 0.32 to 0.53%, indicating that potency of rotogranulated beads can be accurately determined at-line by near-infrared spectroscopy.

It is important to note that although multivariate techniques such as PCR and PLS appear to allow the development of more accurate calibrations, single-wavelength analysis also demonstrated acceptable results. Therefore, after having identified a wavelength region that correlates well with reference method results, it may be possible to employ less expensive and more robust spectrometers for at-line applications.

4.3.2   Wurster coating studies

Wurster coating studies performed by Wargo and Drennen [38] utilized 55% diltiazem HCl beads produced in the laboratory-scale rotogranulation studies. The previously discussed calibration techniques were employed to monitor the coating process and predict the amount of polymer solids applied at various time points during coating. Forty-four samples, collected at various time points during a single 1.8 kg Wurster batch, were used in the calibration development. All calibration models developed yielded similar results. The reported standard error of calibration (SEC) was 0.31% and SEP was 0.23%, demonstrating that the quantity of polymer solids applied during Wurster coating can be evaluated at-line by NIR spectroscopy.

Using pattern recognition algorithms, test samples can be classified as acceptable or unacceptable based on spectral similarity to a training set. In this study, coating samples were classified by a bootstrap pattern recognition technique. Near-infrared spectra from 10 samples, obtained at 16% theoretical applied solids in the first coating trial, were used to develop a spectral training set. During the second coating trial, samples were collected at various time points corresponding to different levels of applied coating. To assess spectral similarity, a distance analogous to a standard deviation is calculated between the test sample and the center of the training cluster. Samples within three standard deviations from the training cluster were considered to contain a desired amount of coating, while samples with distances greater than three standard deviations were classified as outliers. As Figure 4.4 demonstrates, the bootstrap algorithm was effective for identifying when the target 16% coating level was achieved.

More recently, Andersson et al. reported in 1999 the possibility to predict pellet coating thickness with an error as low as 2.2 μm [39]. Lee et al. [40] reported very good agreement between actual measurements and NIR-predicted pellet film thickness for the real-time monitoring and endpoint determination of the coating endpoint.

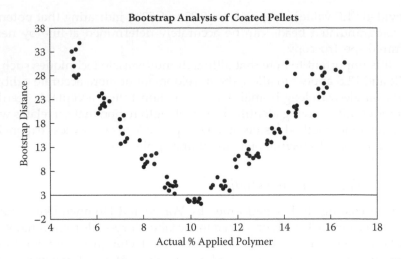

Figure 4.4 Bootstrap standard deviation plot exhibiting the possibility for qualitative identification of polymer film coating endpoint. Dashed line indicates the
three-standard deviation limit for spectral similarity. NIR spectra from 10 samples obtained at the 16% theoretical applied solids level were used as a training group.

References

1. O. Worts and T. Schaefer, Control of Fluidized Bed Granulation V: Factors
   Affecting Granule Growth, *Arch. Pharm. Chem. Sci. Ed.*, 6, 1 (1978).
2. I. A. Chaudry and R. E. King, Migration of Potent Drugs in Wet Granulations,
   J. Pharm. Sci., 61, 1121 (1972).
3. A. M. C. Davies and A. Grant, Air-Conditioning-Generated Noise in a Near-
   Infrared Spectrometer Caused by Fluctuations in Atmospheric Water Vapor,
   Appl. Spectrosc., 41, 7 (1987).
4. J. W. Ellis and J. Bath, Modifications in the Near Infra-Red Absorption Spectra
   of Protein and of Light and Heavy Water Molecules When Water Is Bound to
   Gelatin, *J. Chem. Phys.*, 6, 723 (1938).
5. J. E. Sinsheimer and A. M. Keuhnelian, Near-Infrared Spectroscopy of Amine
   Salts, *J. Pharm. Sci.*, 55(11), 1240 (1966).
6. V. E. Greinacher, W. Luttke, and R. Z. Mecke, The Effect of Temperature upon
   Near Infrared Spectra Absorptivities, *Z. Electrochem.*, 59, 23 (1955).
7. R. J. Warren, J. E. Zarembo, C. W. Chong, and M. J. Robinson, Determination
   of Trace Amounts of Water in Glycerides by Near-Infrared Spectroscopy,
   J. Pharm. Sci., 59(1), 29 (1970).
8. R. G. Whitfield, Near Infrared Reflectance Analysis of Pharmaceutical
   Products, *Pharm. Manuf.*, 3(4), 31 (1986).
9. M. S. Kamat, R. A. Lodder, and P. P. Deluca, Near-Infrared Spectroscopic
   Determination of Residual Moisture in Lyophilized Sucrose through Intact
   Glass Vials, *Pharm. Res.*, 6(11) (1989).

10. I. R. Last and K. A. Prebble, Suitability of Near Infrared Methods for the Determination of Moisture in a Freeze Dried Injection Product Containing Different Amounts of Active Ingredients, *J. Pharm. Biom. Anal.*, 11(11/12), 1071 (1993).

11. J. A. Jones, I. R. Last, B. F. MacDonald, and K. A. Prebble, Development and Transferability of Near-Infrared Methods for Determination of Moisture in a Freeze-Dried Injection Product, *J. Pharm. Biom. Anal.*, 11(11/12), 1227 (1993).

12. W. Plugge and C. van der Vlies, The Use of Near Infrared Spectroscopy in the Quality Control Laboratory of the Pharmaceutical Industry, *J. Pharm. Biom. Anal.*, 10(10/12), 797 (1992).

13. W. Plugge and C. van der Vlies, Near-Infrared Spectroscopy as an Alternative to Assess Compliance of Ampicillin Trihydrate with Compendial Specifications, *J. Pharm. Biom. Anal.*, 11(6), 435 (1993).

14. Software manual, issued by NIRSystems, *IQ2 Version 1.02 Software Manual*, NIR Systems, Silver Spring, MD, 1990.

15. P.J. Gemperline, L.D. Webber, and F.O. Cox, Raw Materials Testing Using Soft Independent Modeling of Class Analogy Analysis of Near-Infrared Reflectance Spectra, *Analyt. Chem.*, 61, 138–144 (1989).

16. J. G. White, Online Moisture Detection for a Microwave Vacuum Dryer, *Pharm. Res.*, 11(5), 728 (1994).

17. P. P. Sanghvi, J. K. Drennen, and C. C. Collins, Discriminant Analysis of Pharmaceutical Components Using NIRA, presented at AAPS Annual Meeting, San Diego, CA, November 1994, paper PT 6058.

18. S. H. Han and C. K. Oh, Analysis of TLC Spots via NIRA, presented at AAPS Annual Meeting, Seattle, WA, October 1996, paper APQ 1176.

19. P. Frake, D. Greenhalgh, S. M. Grierson, J. M. Hempenstall, and D. R. Rudd, Process Control and Endpoint Determination of a Fluid Bed Granulation by Application of Near-Infrared Spectroscopy, *Int. J. Pharm.*, 151(1), 75 (1997).

20. J. Rantanen, S. Lehtola, P. Ramet, J.-P. Mannermaa, and J. Yliruusi, On-Line Monitoring of Moisture Content in an Instrumented Fluidized Bed Granulator with a Multi-Channel NIR Moisture Sensor, *Powder Technol.*, 99, 163 (1998).

21. E. W. Ciurczak, R. P. Torlini, and M. P. Demkowicz, Determination of Particle Size of Pharmaceutical Raw Materials Using Near Infrared Reflectance Spectroscopy, *Spectroscopy*, 1, 36 (1986).

22. J. Rantanen and J. Yliruusi, Determination of Particle Size in a Fluidized Bed Granulator with a Near Infrared Set-Up, *Pharm. Pharmacol. Commun.*, 4, 73 (1998).

23. F. J. S. Nieuwmeyer, M. Damen, A. Gerich, F. Rusmini, K. van der Voort Maarschalk, and H. Vromans, Granule Characterization during Fluid Bed Drying by Development of a Near Infrared Method to Determine Water Content and Median Granule Size, *Pharm. Res.*, 24(10), 1854 (2007).

24. M. Alcala, M. Blanco, M. Bautista, and J. M. Gonzalez, On-Line Monitoring of a Granulation Process by NIR Spectroscopy, *J. Pharm. Sci.*, 99(1), 336 (2010).

25. L. Chablani, M. K. Taylor, A. Mehrotra, P. Rameas, and W. C. Stagner, Inline Real-Time Near-Infrared Granule Moisture Measurements of a Continuous Granulation–Drying–Milling Process, *AAPS PharmSciTech*, 12(4), 1050 (2011).

26. G. X. Zhou, Z. Ge, J. Dorwart, B. Izzo, J. Kukura, G. Bicker, and J. Wyvratt, Determination and Differentiation of Surface and Bond Water in Drug Substances by Near Infrared Spectroscopy, *J. Pharm. Sci.*, 92(5), 1058 (2003).

27. W. Li, J. Cunningham, H. Rasmussen, and D. Winstead, A Qualitative Method for Monitoring of Nucleation and Granule Growth in Fluid Bed Wet Granulation by Reflectance Near-Infrared Spectroscopy, *J. Pharm. Sci.*, 96(12), 3470 (2007).

28. A. Miwa and K. Makado, A Method for Predicting the Amount of Water Required for Wet Granulation Using NIR, *Int. J. Pharm.*, 376, 41 (2009).

29. A. Miwa, T. Yajima, and S. Itai, Prediction of Suitable Amount of Water Addition for Wet Granulation, Determination of Moisture in Pharmaceutical Granulations, *Int. J. Pharm.*, 195, 81 (2000).

30. A. Miwa, T. Yajima, H. Ikuta, and K. Makado, Prediction of Suitable Amounts of Water in Fluidized Bed Granulation of Pharmaceutical Formulations Using Corresponding Values of Components, *Int. J. Pharm.*, 352, 202 (2008).

31. A. Peinado, J. Hammond, and A. Scott, Development, Validation and Transfer of a Near Infrared Method to Determine In-Line the End Point of a Fluidised Drying Process for Commercial Production Batches of an Approved Oral Solid Dose Pharmaceutical Product, *J. Pharm. Biomed. Anal.*, 13 (2011).

32. M. Andersson, M. Josefson, F. W. Langkilde, and K.-G. Wahlund, Monitoring of a Film Coating Process for Tablets Using Near Infrared Reflectance Spectrometry, *J. Pharm. Biomed. Anal.*, 20, 27 (1999).

33. J. J. Moes, M. M. Ruijke, E. Gout, H. W. Frijlink, and M. I. Ugwoke, Application of Process Analytical Technology in Tablet Process Development Using NIR Spectroscopy: Blend Uniformity, Content Uniformity and Coating Thickness, *Int. J. Pharm.*, 357, 108 (2008).

34. C. Gendre, M. Gent, M. Boire, M. Julien, L. Meunier, O. Lecoq, M. Baron, P. Chaminade, and J. M. Péan, Development of a Process Analytical Technology (PAT) for In-Line Monitoring of Film Thickness and Mass of Coating Materials during a Pan Coating Operation, *Eur. J. Pharm. Sci.*, 43, 244 (2011).

35. C.-V. Möltgen, T. Puchert, J. C. Menezes, D. Lochmann, and G. Reich, A Novel In-Line NIR Spectroscopy Application for the Monitoring of Tablet Film Coating in an Industrial Scale Process, *Talanta*, 92, 26 (2012).

36. M.-J. Lee, C.-R. Park, A.-Y. Kim, B.-S. Kwon, K.-H. Bang, Y.-S. Cho, M.-Y. Jeong, and G.-J. Choi, Dynamic Calibration for the In-Line NIR Monitoring of Film Thickness of Pharmaceutical Tablets Processed in a Fluid-Bed Coater, *J. Pharm. Sci.*, 99(1), 325 (2010).

37. Y. Kato, D. Sasakura, T. Miura, A. Nagatomo, and K. Terada, Evaluation of Risk and Benefit in the Application of Near-Infrared Spectroscopy to Monitor the Granule Coating Process, *Pharm. Dev. Technol.*, 13, 205 (2008).

38. D. J. Wargo and J. K. Drennen, presented at AAPS Annual Meeting, Seattle, WA, October 1996, paper PT 6116.

39. M. Andersson, S. Folestad, J. Gottfries, M. O. Johansson, M. Josefson, and K.-G. Wahlund, Quantitative Analysis of Film Coating in a Fluidized Bed Process by In-Line NIR Spectrometry and Multivariate Batch Calibration, *Anal. Chem.*, 72, 2099 (2000).

40. M. J. Lee, D.-Y. Seo, H.-E. Lee, I.-C. Wang, W.-S. Kim, M.-Y. Jeong, and G. J. Choi, In Line NIR Quantification of Film Thickness on Pharmaceutical Pellets during a Fluid Bed Coating Process, *Int. J. Pharm.*, 403, 66 (2011).

chapter 5

Pharmaceutical assays

5.1 Introduction

Although first reported by Herschel in 1800, the near-infrared (NIR) region was ignored until the late 1950s. Publications describing pharmaceutical applications appeared approximately 10 years later, with the majority appearing since 1986. Reviews of NIR spectroscopy were published in the early 1990s [1, 2] and contain references to earlier reviews. Ciurczak published a comprehensive review of pharmaceutical applications [3]. More recent reviews cover the various topics that are now studied by NIR [4–10]. Several texts on NIR are also available [11–16], several of which contain chapters on pharmaceutical applications.

The conventional NIR region lies between 700 and 2500 nm. Near-infrared spectra arise from absorption bands resulting from overtones and combinations of fundamental mid-IR (MIR) stretching and bending modes. They have low molar absorptivity with broad, overlapping peaks. The low absorptivity is a primary reason for the usefulness of the method for analysis of intact dosage forms. These absorbance arise from C–H, O–H, and N–H bonds.

The earliest publications of NIR assays of pharmaceuticals appeared in the late 1960s, although these first pharmaceutical applications did not involve intact dosage forms. Usually the drug was extracted, then analyzed. In some cases, solid-state spectra were collected from ground dosage forms.

In 1966, Sinsheimer and Keuhnelian investigated a number of pharmacologically active amine salts both in solution and in the solid state [17]. In 1967, Oi and Inaba quantified two drugs: allylisopropylacetureide (AL) and phenacetin (PH) [18]. Samples were dissolved in chloroform and quantified at 1983 nm for AL and 2019 nm for PH.

Sinsheimer and Poswalk determined water in several matrices [19]. Solid samples were analyzed for hydrous and anhydrous forms of strychnine sulfate, sodium tartrate, and ammonium oxalate mixed with KCl and compressed into disks containing 100 mg KCl and 25 mg drug. The water band at 1940 nm was seen in the hydrates for some samples.

5.2   Qualitative analysis

5.2.1   Raw materials

A landmark paper presented by John Rose in 1982 [20] showed that a large number of structurally similar penicillin-type drugs could be identified and determined using NIR techniques developed at his company.

In 1984, Mark introduced the Mahalanobis distance in an algorithm for discriminant analysis of raw materials. The theory behind the software was described in a paper by Mark and Tunnell [21] and first applied to pharmaceuticals by Ciurczak [22]. With the advent of 100% testing of incoming raw materials, qualitative analysis of raw materials by NIR became popular quickly.

For qualitative analysis using pure materials and pure liquids, or where few samples exist, the discriminant technique may prove difficult; little variation exists and a new, perfectly acceptable sample may be misidentified due to small differences not seen in the original sample set. Ciurczak [23] suggested a technique where artificial samples may be made either physically or electronically. Ciurczak also reported on the use of spectral matching and principal component evaluation for raw materials testing [24, 25], as well as components of granulations or blending studies [26, 27]. In-process blend uniformity testing may make such qualitative testing valuable for all products manufactured in the United States [28].

Rodionova et al. used principal component analysis (PCA) and SIMCA to determine the identification of raw materials in their original packaging [29]. Chen et al. [30] discuss the development of a multisite multi-instrument database used for the identification of raw materials. Authors provided insights regarding the validation and routine testing methods. More recently, Wen et al. [31] compared several spectroscopic techniques and determined that NIR was particularly well suited for detecting the presence of adulterants but not their identity.

5.2.2   Blending studies

Because almost all materials used in the pharmaceutical industry have NIR spectra, the use of NIR for assuring blend homogeneity may prove to be a valuable application. Ciurczak [25, 32] reported some of the first work on this subject. His work involved the use of a fiber probe to collect spectra from various locations in the mixer. Spectral matching and PCA were used to measure how similar the powder mix in a particular portion of the blender was to a predetermined good, or complete, mix. The match index or PCA scores were plotted versus time to assess the optimal blending time.

Drennen, in his thesis defense, first used NIR to study homogeneity during processing of semisolid dosage forms [33]. The translucent state of the samples was addressed in the paper. Later he used similar approaches for the study of powder blend homogeneity. Kirsch and Drennen covered the state of the art in 1995 [34].

Since then, numerous qualitative applications for blend monitoring have been developed. Chapter 3 of this book presents in detail these studies. Principal component analysis has primarily been the tool of choice [35–37] because of its ability to represent the variance of the spectral data with a limited number of variables. More simplistic spectral comparison methods based on standard deviations were also investigated [38].

5.2.3 Verification of supplies for double-blind clinical studies

Clinical trials usually involve the use of placebos that are deliberately made to look like the actual dosage form. Often various levels of actives are present, again in dosage forms that appear identical. The most common approach for identifying the blinded samples has been to sacrifice some of the blister packs to ascertain whether the materials are in the correct order. Alternatively, tablets or other solid dosage forms may be analyzed directly through the clear polymer casing of the blister packs in a noninvasive and nondestructive NIR assay for identification of placebos and samples with different dosage levels.

In two papers, Ritchie [39, 40] described an approach for performing qualitative NIR analysis of clinical samples with a concern for current good manufacturing practices (cGMPs). Since clinical lots are often ad hoc formulations, it is difficult to generate a discriminant equation prior to the actual clinical trial. Ritchie et al. developed a procedure whereby equations are quickly generated for any particular study, then discarded.

Dempster et al. used three sampling configurations to qualify clinical samples of tablets containing 2, 5, 10, and 20% concentrations of active agent, plus a matching placebo and a marketed drug used as clinical comparators [41]. The first configuration required that the tablets be removed from the blister packs. In the second, tablets were scanned through the plastic packaging using a reflectance module. With the third arrangement, tablets were analyzed through the plastic blister packaging with a fiber optic probe.

Using the first configuration, all but the 2% tablets were easily classified; the 2% tablets could not be differentiated from placebo samples. With the second and third configurations, only 10 and 20% tablets, placebo, and clinical comparator tablets could be properly classified.

Another application of NIR in the analysis of clinical batches was published in 1994 by Aldridge et al. [42]. A NIRSystems Model 6500 spectrometer with a custom sampling configuration was used for spectral collection from the blister-packed samples. In 1998, Candolfi et al. [43] combined pattern recognition techniques with NIR spectroscopy to identify samples of clinical study lots. Authors used linear discriminant analysis (LDA), quadratic discriminant analysis, and K-nearest neighbors. Dimensionality reduction by PCA followed by LDA gave the best results for the classification of capsules and tablets based on their concentration in active. De Maesschalck et al. [44] used a classification method based on PLS to confirm the identity of double-blind clinical trial tablets.

5.2.4   Active ingredients within dosage forms

A 1986 paper by Ciurczak and Maldacker [45] using NIR for tablet formulation blends examined the use of spectral subtraction, spectral reconstruction, and discriminant analysis for the analysis of dosage forms. Blends were prepared where actives—aspirin (ASA), butalbital (BUT), and caffeine (CAF)—were omitted from the formulation or varied over a range from 90 to 110% of label strength. For spectral subtraction, spectra of true placebos were subtracted, yielding spectra very close to those of the omitted drug.

Identification of constituents by spectral reconstruction was performed with commercially available software, based on work by Honigs [46], later expanded upon by Honigs et al. [47]. Using a series of mixtures of known concentrations, the spectrum of the drug was reconstructed, providing identification of actives in the blend.

A third set of experiments classified samples by discriminant analysis. In one series of blends, the CAF, BUT, and ASA concentrations varied independently between 90 and 110% of the labeled claim. In another series, one of the three drugs was excluded from the mixture, while the others were varied between 90 and 110%. The Mahalanobis distance statistic was used for classification of formulations. This technique was used for samples of complete formulations (all three drugs at 100% of label strength), borderline formulations, and samples lacking one active component.

In 1986, Whitfield [48] used discriminant analysis to ascertain that a veterinary drug dosed in chicken feed was present before conducting a quantitative analysis.

A considerable amount of (unpublished) work has been performed by Ciurczak on counterfeit tablets. Using the same algorithms that have been applied to discriminate between placebos and active products, counterfeit products may be easily identified. The spectral variability stems from different raw materials and manufacturing processes, even though the active may be present at the correct level. Moffat et al. [49], Rodionova et al. [50], and Polli et al. [51] published examples of use of NIR to detect counterfeit drugs.

Said et al. [52] used PCA and SIMCA to compare paracetamol tablets made in the UK and in Malaysia and to evaluate the manufacturing variability existing between the two countries. They proposed the approach as a detection method to detect substandard pharmaceutical products. In 2012, Shi et al. proposed a qualitative method to detect tablets presenting extremely high or low concentrations [53]. Authors proved that when converting sample concentration into scores, it was possible to establish the content uniformity of a batch that is comparable with more traditional random sampling followed by concentration determination by high-performance liquid chromatography (HPLC).

5.2.5 Packaging materials

As indicated in Chapter 4 in this text, polymers have been analyzed by NIR for some time. In 1985, Shintani-Young and Ciurczak [54] used discriminant analysis to identify polymeric materials used in packaging: plastic bottles, blister packaging, and PVC wrap, to name a few. Replacement of the time-intensive infrared spectroscopy, using attenuated total reflectance (ATR) cells, was considered quite good. Information such as density, cross-linking, and crystallinity may be measured. Near-infrared testing of polymeric materials is covered in detail by Kradjel and McDermott [55]. Laasonen et al. [56] used NIR to determine the thickness of plastic sheets used in blister packaging.

5.2.6 Polymorphism

When organic (drug) molecules crystallize from a solvent, the crystal structure is dependent upon the speed of crystallization, temperature, polarity of the solvent, concentration of the material, etc. Since the energy of the crystal affects the (physiological) rate of dissolution, and thus the potency and activity of the drug, polymorphism is an important pharmaceutical concern [57]. The most common tool to determine crystal form is differential scanning calorimetry (DSC). Unfortunately, DSC uses small samples and may not represent the bulk of the material. X-ray diffraction is another excellent technique, but quite slow and sometimes difficult to interpret.

In 1985, Ciurczak [58] reported using NIR to distinguish between the polymorphic forms of caffeine. Polymorphism was also reported upon by Gimet and Luong [59] in 1987. They found NIR a useful tool to ascertain whether the processing of a granulation led to any crystallinity changes of the active material. It has been noted in the past that physical processes such as milling, wet granulating, or compressing of tablets can cause polymorphic changes of a drug substance.

Miller and Honigs showed in 1985 that the small difference in position of the C-H overtone and combination bands existing between α and γ crystalline forms of glycine could be observed by NIR [60]. Dreassi et al. [61] showed that NIR was able to distinguish between two forms of gemfibrozil that had recrystallized in two different solvents, and that four polymorphs of chemodeoxycholic acid were easily separable using Mahalanobis distance.

Aldridge et al. [62] used pattern recognition to differentiate between the desired and unwanted polymorphs of an active substance. More importantly, the method was transferred to at least six other instruments for application. Polymorphism of drug substances was studied by De Braekeleer et al. [63] in 1998. They used PCA, SIMPLISMA, and orthogonal projections to correct for temperature variation during the monitoring of polymorph conversion. This was performed in real time, online in a commercial process.

Near infrared has been used to quantify amorphous content in a crystalline matrix [64], mixtures of multiple polymorphic forms [65], and physical mixtures of crystalline and amorphous drug in the presence of excipients [66, 67]. Otsuka et al. [68] found that Fourier transform (FT)–NIR outperformed x-ray powder diffraction (XRPD) with respect to accuracy statistics over the calibration range and down to 1% (w/w). Other quantitative studies have reported prediction errors of 5 to 6% (w/w) of γ-IMC (indomethacin) in a sample matrix including amorphous and α-IMC [69]. These results demonstrate NIR selectivity of the solid-state forms of IMC. Selectivity, the proportion of analyte signal unaffected by other spectral interferences, is especially critical for monitoring dispersions where physical instability could manifest in a combination of amorphous and crystalline forms (including various polymorphs). Similar studies were performed for lactose and showed the possibility to quantify down to 0.5% of crystalline material in an amorphous base [70].

Near-infrared spectroscopy has been used as a process analytical technology to monitor polymophic and pseudopolymorphic changes during manufacturing. The hydration of theoplylline anhydrous was monitored during wet granulation by Räsänen et al. [71]. Authors were able to detect different states of water molecules during the wet granulation process faster and in a more flexible manner than with conventional methods. Davis et al. [72] monitored the polymorphic transformation of glycine during the drying phase of wet granulation. Pseudopolymorph transformation of nitrofurantoin and theophylline during pellet manufacturing by extrusion spheronization was also investigated [73]. The formation of ibuprofen-nicotinamide co-crystals by solvent-free co-crystallization was monitored by NIR, directly in the screw extruder die [74].

5.2.7   Optical isomers

In a presentation by Ciurczak [75], it was observed that pure *d-* and *l-*amino acids gave identical NIR spectra, while racemic crystals generated quite different spectra. A 1986 paper outlines work later completed by Buchanan and colleagues [76, 77]. In this work, varying percentages of *d*- and *l*-valine were mixed physically and scanned by NIR. The spectra were identical except for particle size-induced baseline shifts. These mixtures were then dissolved and recrystallized as racemic crystals. These new samples were scanned by NIR; obvious qualitative and quantitative differences were observed. Dreassi et al. [61] showed that NIR was able to distinguish between levamisole and tetramisole (racemic mixture).

Mustillo and Ciurczak [78] presented a paper discussing the spectral effect of optically active solvents on enantiomers. This information was later used to screen for polar modifiers in normal phase chromatographic systems where racemic mixtures were involved [79].

5.2.8   Structural isomers

Structural or geometric isomers may be distinguished by NIR. The xanthines (caffeine, thobromine, and theophylline) were discriminated as shown in a paper by Ciurczak and Kradjel [80]. In that same presentation, ephedrine and pseudoephedrine were shown to have different spectra. The difference in many of these cases is one methyl group or the exchange of position of a (–H) with a (–OH) on the same carbon atom. The sensitivity of NIR to intermolecular bonding makes the technique valuable for detecting such position exchanges.

5.3   Quantitative analysis

5.3.1   Particle size

It has long been recognized that the spectra of powdered samples are affected by their particle sizes [81, 82]. The effect of particle size difference is usually seen as a sloping baseline in the spectrum, where baseline increases at longer wavelengths. Many approaches were suggested to circumvent this problem, including physical (screening, grinding) and mathematical (second-derivative, multiplicative scatter correction) corrections. In 1985, Ciurczak et al. [83, 84] presented work showing that there was a linear relationship between the absorbance at any wavelength and the reciprocal of the particle size. The calibration for the project was by low-angle laser scattering (LALS).

Ilari et al. [85] used scatter correction to determine the particle sizes of materials. Particle sizes of both organic and inorganic materials were determined by this technique. O'Neil et al. [86] measured the cumulative particle size distribution of microcrystalline cellulose with diffuse reflectance NIR. Both multiple linear regression (MLR) and PCA were used for the work. The results were consistent with those obtained by forward-angle laser light scattering. Rantanen and Yliruusi [87] predicted granule particle size by NIR during granulation. Similar results were obtained by Nieuwmeyer et al. [88] and Otsuka [89], who were able to develop prediction models using partial least-squares for the prediction of median granule size.

5.3.2   Moisture

Because water provides the greatest extinction coefficient in the NIR for pharmaceutically relevant materials, it stands to reason that this is one of the most measured substances by the NIR technique. Derksen et al. [90], for instance, used NIR to determine water through the moisture content of samples with varying active content. Kamet et al. applied NIR to moisture determination of lyophilized pharmaceutical products [91]. Brülls et al. [92] reported a good agreement between "NIRS monitoring and product temperature monitoring about the freezing process and the transition from frozen solution to ice-free material." Authors reported an increased understanding of the phenomena taking place during the process, such as the rate of the desorption process and the steady state where the drying was complete.

Izutsu et al. used NIR to detect the presence of interactions between the amino or guanidyl groups of L-arginine and phosphate ions in amorphous freeze-dried cakes [93]. The quantification of surface and bond water for lyophilized mannitol was investigated [94]. Authors found the prediction error for hydrate water (SEP = 0.50%) to be almost twice as large as that for the prediction of surface water (SEP = 0.22%), but determined that NIR was particularly well suited to control pseudopolymorphic transformation during lyophilization.

Warren et al. [95] described a technique for determining water in glycerides. Transmission spectra of propylene glycol and glycerine were used to calibrate and measure the water content. Correlation of total, bound, and surface water in raw materials was the subject of a paper by Torlini and Ciurczak [96]. In this paper, NIR was calibrated by Karl Fischer titration, differential scanning calorimetry (DSC), and thermogravimetric analysis. It was seen that there was a qualitative difference between surface and bound water that could be distinguished by NIR, but not by chemical or typical loss on drying (LOD) techniques. The thermal analysis methods were needed for calibration. Zhou et al. [97] discussed the capacity of using principal component analysis to differentiate between the removal of surface water and bond water and the possibility to determine precise granulation endpoints in situations where several hydrates coexist.

5.3.3   Hardness

One quality control application of near-infrared spectroscopy is the non-destructive determination of tablet hardness. Near-infrared prediction of tablet hardness has been used and investigated by Drennen for a number of years. In 1991, the first publication of the NIR technique for this application appeared [25]. Ciurczak and Drennen published similar results in 1992 [98], and Drennen and Lodder in 1993 [99]. Results of a study presented at the 1994 annual meeting of the American Association of Pharmaceutical Scientists were published in a 1995 paper [100]. In that paper, Kirsch and Drennen identified the utility of the technique in the determination of multiple film-coated tablet properties, including tablet hardness. In a review paper regarding the use of near infrared in the analysis of solid dosage forms, Kirsch and Drennen discussed the historical aspects of near-infrared prediction of tablet hardness [101].

Changes in dosage form hardness are seen as sloping spectral baseline shifts in which the absorbance increases as hardness increases (Figure 5.1). This is more pronounced at longer wavelengths where most samples have a higher baseline absorbance and is due to a multiplicative light scattering effect [102]. This is the most obvious spectral change seen with increasing tablet hardness, although for some samples other spectral changes occur [103].

To simplify hardness determination, Kirsch and Drennen [104] presented a new approach to tablet hardness determination that relies on simpler, more understandable statistical methods and provides the essence of the full spectral multivariate methods, but does not depend upon

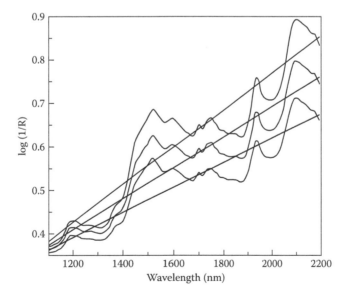

Figure 5.1 Spectra of 20% cimetidine tablets with hardness varying from 1.5 kp (bottom) to 6.5 kp (top). Note the increasing baseline and increasing slope as hardness increases.

individual wavelengths of observation. Specifically, the previous hypotheses regarding the spectroscopic phenomena that permit NIR-based hardness determinations were examined in greater detail. Additionally, a robust method employing simple statistics based upon the calculation of a spectral best fit was developed to permit NIR tablet hardness prediction across a range of drug concentrations.

The character of a spectrum, its pattern of peaks and valleys due to the chemistry of a sample, varies little with changes in tablet hardness. However, as described, a significant shift in the spectral baseline of a tablet occurs with increasing hardness due to the physical changes brought about by increasing compacting force. The proposed approach exploits this baseline shift and involves the determination of a best-fit line through each spectrum, thereby reducing the spectrum to slope and intercept values. This provides two advantages. First, the calculation of the best-fit line through the spectrum characterizes this change in spectral baseline slope. Second, the entire spectrum can be used, which means that the absorbance of any one peak or band does not unduly change the slope of the regression line. Therefore, variations in formulation composition are averaged across the entire spectrum and are less likely to affect the slope of the best-fit line.

After the calculation of the best-fit line, each near-infrared spectrum has been reduced to slope and intercept values. Using ordinary least-squares, a calibration can be developed that regresses the slopes and intercepts of the best-fit lines against the laboratory-determined hardness values of the respective tablet samples. The prediction of an unknown tablet's hardness involves the collection of a near-infrared spectrum, reduction of that spectrum to slope and intercept values, and subsequent determination of the tablet's hardness from the calibration equation (Figure 5.1). The advantages of this approach are its deweighting of individual absorbance peaks and valleys, its use of simple and understandable statistics, and the lack of a need to make judgment calls concerning the inclusion or exclusion of latent variables during calibration development.

Additionally, this robust method employing simple statistics based upon the calculation of a spectral best fit permits near-infrared tablet hardness prediction across a range of drug concentrations. In 1997, Morisseau and Rhodes [105] published a paper similar to the work by Drennen and Kirsch wherein they used NIR to determine the hardness of tablets. Four formulations (two of hydrochlorothiazide (HCTZ) and two of chlorpheniramine (CTM)) and a placebo were prepared with hardness levels between 2 and 12 kg. Using MLR and partial least-squares (PLS), equations were generated that allowed good prediction of hardness for all the products.

Numerous additional authors have contributed to the study of hardness measurement by NIR [106–112]. Chen et al. [107] found that artificial neural networks results were comparable to PLS predictions, and Donoso et al. [108] compared several regression techniques and determined that PLS was performing the best.

More recently, the distribution of density within tablets was investigated with chemical imaging. In 2008, Ellison et al. [113] found that the intratablet density uniformity was strongly dependent on the friction between the powder and die walls and showed that tablets with no magnesium stearate or 0.25% magnesium stearate were less uniform than tablets with 1.0% magnesium stearate. In 2011, Igne et al. [114] determined that the distribution of predicted density values could be used to determine the radial tensile strength of tablets not only right after compression, but also after relaxation, thus permitting the development of a method to determine in real time, at the tablet press, the final tablet hardness.

5.4 Determination of actives in tablets and capsules

In the earliest NIR assays, tablets and capsules were not analyzed intact. Before NIR spectral collection, drugs were extracted from the matrix into solution. The first reported use of NIR for tablets was by Sherken in 1968 [115]. In this study, meprobamate in tablet mixtures and commercially available preparations was assayed. Two wavelengths, corresponding to the symmetric and asymmetric stretching modes of the primary amine group in the drug molecule, were used.

Allen used NIR for the quantitative determination of carisoprodol, phenacetin, and caffeine [116]. Twenty tablets were pulverized and an aliquot dissolved in chloroform was measured. Standard solutions of carisoprodol, phenacetin, and caffeine were scanned between 2750 and 3000 nm. Caffeine and phenacetin were determined at 2820 nm (carisoprodol) and 2910 nm (phenacetin), with caffeine determined at 3390 nm. The coefficient of variation (CV) was 1.4% or less.

In 1977, Zappala and Post used NIR for meprobamate in four pharmaceutical preparations: tablets, sustained-release capsules, suspensions, and injectables [117]. The NIR method was an improvement over that introduced by Sherken; it took advantage of a meprobamate (primary amine) combination band at 1958 nm, not subject to the interference suffered by the peak at 2915 nm. Twenty tablets or capsules were pulverized, and an aliquot dissolved in chloroform was scanned. Nine commercial products from four manufacturers were analyzed. The coefficient of variation was 0.7% for tablets and 1.3% for capsules (1.5% for the reference method).

In 1990, Corti et al. used an extraction prior to NIR to improve the detection limit [118]. Oral contraceptives were used in the study for ethinylestradiol (ETH) and norethisterone, two synthetic hormones. Qualitative and quantitative analyses were desired. Tablets of 80 mg (0.05 mg ethinylestradiol and 0.25 mg norethisterone) were extracted with chloroform and scanned. In a semiquantitative method, six wavelengths were used in a Mahalanobis distance calculation, and it was possible to distinguish the ETH extracts at concentrations below 0.05%. For quantitative analysis, multiple linear regression was employed. The coefficients of determination obtained were $r^2 = 0.85$ (ethinylestradiol) and $r^2 = 0.86$ (norethisterone). With low drug concentrations and a short range of values, the SECs were high.

Near-infrared spectroscopy proved valuable for the analysis of pharmaceutical powders in a 1981 paper by Becconsall et al. [119]. Near-infrared and UV photoacoustic spectroscopy were used for determination of propranolol–magnesium carbonate mixtures. Spectra were collected from 1300 to 2600 nm with carbon black as the reference. An aromatic C–H combination band at 2200 nm and an overtone band at 1720 nm were used to quantify propranolol. In this case, the UV data were nonlinear, while the NIR method provided a linear calibration.

In 1982, Ciurczak and Torlini published on the analysis of solid and liquid dosage forms [120]. They contrasted NIR calibrations for natural products versus those for pharmaceuticals. Samples prepared in the laboratory are spectrally different from production samples. Using laboratory samples for calibration may lead to unsatisfactory results; production samples for calibration are preferred for calibration. Near infrared was compared with HPLC for speed and accuracy. The effect of milling the samples prior to analysis was also investigated. Two dosage form matrices were studied: a caffeine plus acetaminophen mixture and an acetaminophen mixture. The acetaminophen mixtures were analyzed after milling, and caffeine-acetaminophen mixtures were analyzed with and without milling. Multiple linear regression was used for the calibration.

Milling of the caffeine-acetaminophen mixture improved the determination of acetaminophen, but caffeine was unchanged. The difference between theoretical and predicted concentrations was ~0.25%, competitive with HPLC. Near infrared allowed rapid analysis times with no expense for solvent purchase and disposal.

In 1987, Chasseur assayed cimetidine granules [121]. Individual batches of granules were prepared with cimetidine concentrations ranging from 70 to 130% of labeled potency. For calibration, first- and second-derivative spectra and one or two wavelengths were included in the model. A two-wavelength model using first-derivative spectra gave optimal results, with SEP = 1.75%. The SEP was 2.73% for the NIR and 2.97% for UV.

A 1987 paper by Osborne used NIR to determine nicotinamide in vitamin premixes [122]. High-performance liquid chromatography, the reference method for nicotinamide (NIC), required 3 days to analyze 36 samples; the NIR method required only 30 min.

Twenty-five mixtures were used for calibration, with concentrations from 0 to 6%. Spectra were collected between 1200 and 2400 nm. Second-derivative spectra were calculated, and the calibration obtained the ratio of the second-derivative values at 2138 nm (nicotinamide) and 2070 nm (a spectral minimum). The SEP for the validation set was 0.56% w/w. Both HPLC and NIR gave comparable results.

In a 1988 paper, Lodder and Hieftje used the quantile-BEAST (bootstrap error-adjusted single-sample technique) [123] to assess powder blends. In the study, four benzoic acid derivatives and mixtures were analyzed. The active varied between 0 and 25%. The individual benzoic acid derivatives were classified into clusters using the standard deviations (SDs). Acetylsalicylic acid was added to the formulations at concentrations of 1 to 20%. All uncontaminated samples were correctly identified. Simulated solid dosage forms containing ratios of the two polymorphs were prepared. They were scanned from 1100 to 2500 nm. The CVs ranged from 0.1 to 0.9%.

Near infrared was used in 1989 to quantify ketoprofen in gel and powder matrices for encapsulation [124]. Two ranges were used: ± 5% of theory and 3 to 30% active. The SEP was approximately 2%, with no sample having an error greater than 3.5%.

Corti et al. analyzed ranitidine and water in sample tablets [125]. Production samples, when made under control, contain a narrow range of values for active concentration. It is then difficult to cull any number of samples to generate a desired range of sample values in the calibration set to cover a 20% (90 to 110% of label claim) range as done for typical HPLC methods. The result is a diminished correlation coefficient due to the small range of values in the calibration set. Actual drug content of the samples was determined by HPLC; water content was determined by Karl Fischer moisture analysis. For prediction of the drug content, three NIR calibrations using MLR were developed. Lab samples provided a SEP of 8.4% with unknown samples. The second calibration of production samples provided a SEP of 1% for production samples and 6.4% for lab samples. A third calibration, using both production and modified samples, gave SEPs of ~1% for both.

The optimal calibration provided a range of ~5%. The calibration for water employed production samples as is and modified with additional moisture (vapor phase addition). The SEP was below 0.1%. For production samples over a 1-year period, the NIR method had the greatest error for moisture with SEP under 1%. As a qualitative test, it erroneously rejected samples with a moisture content of >2%. The results showed that, for products with little production variability, a small number of samples (~10 to 20) are sufficient for an initial equation. This may be built upon with batches containing less or more than the amounts used for the initial calibration.

In 1991, Ryan et al. [126] reported one study where NIR was found to be unsuitable for its intended application. The purpose was to find a rapid method for the verification of clinical packaging. Both MIR and NIR were used to identify two cholesterol-lowering drugs: lovastatin and simvastatin. Both methods provided detection limits of ~1% (w/w) for ground samples. Near infrared was unable to differentiate the two drugs at concentrations below 0.1%. In 1992, Corti et al. analyzed antibiotic compounds by NIR [127]. Multiple linear regression (MLR) was used for the quantitation and Mahalanobis distances for qualitative analysis. Qualitative analysis (using Mahalanobis distances) differentiated 10 antibiotic preparations, including three types of ampicillin and blends of erythromycin powder and granules. The SEP for each calibration was below 2%.

A 1993 paper by Blanco et al. addressed concerns of laboratory manipulation of production samples prior to analysis [128]. Two commercial preparations of ascorbic acid (vitamin C) were analyzed, including one granular product and one effervescent tablet. No less than five batches were used. All samples were ground to a specific mesh size (either 250 or 100 μm). To expand the calibration range, samples were diluted with the primary filler or overdosed with ascorbic acid. Prior to analysis, three preprocessing methods were evaluated: multiplicative scatter correction (MSC), signal scaling, and first derivative. In this study, first-derivative spectra provided the best calibration results.

Stepwise MLR calibrations used up to four wavelengths and provided r values higher than 0.99 and SEPs below 2.4%. Calibrations were developed using PLS on full- (1100 to 2500 nm) and reduced-wavelength (1300 to 1800 nm) ranges. Two or three factors were adequate for the calibrations, giving SEP values under 2%. Stepwise MLR was more accurate for the simpler granule preparation, while PLS was more accurate for the effervescent formulation.

From these early developments, NIR spectroscopy has been used extensively in pharmaceutical applications. Very good and exhaustive reviews by Forina et al. [4], Reich [5], Roggo et al. [6], and Luypaert et al. [7] summarize the recent development in pharmaceutical assays and more particularly active pharmaceutical ingredient (API) quantification. It is, however, necessary to briefly emphasize several challenging applications that have been successfully completed. The prediction of lubricant (magnesium stearate) was found to be possible at levels as low as 0.5% w/w [129]. Ito et al. [130] showed the feasibility to use NIR to predict the contact of API in intact bilayer tablets, and the research team showed that it was unnecessary to control the side that was irradiated when working in transmittance configuration. The prediction of API through coating was found possible by Boiret et al. [131].

5.5   Considerations for intact dosage form analysis

Numerous styles and brands of instruments and sample cells have been used for the analysis of tablets. The authors currently use several brands of instrumentation for tablet analysis, including filter-, diffraction grating-, and acousto-optic tunable filter-based instrumentation. Detector configurations are evolving slowly toward an optimal design; however, the designs of most manufacturers are suitable for many applications. Tablets have been successfully analyzed with integrating spheres and with a standard dual-angled detector configuration. Intact tablets are analyzed in both diffuse reflectance and transmittance modes. Chapter 2 is dedicated to instrumentation.

The first analyses of individual intact tablets and capsules involved the use of reflective aluminum sample cells, designed specifically for tablets or capsules, allowing illumination of all surfaces and probably collecting transmitted signal. Illumination of all surfaces enhances sensitivity.

Sample positioning is the single largest source of error in NIR analysis of tablets in diffuse reflectance or transmission measurements. Hardware, methodology, and mathematics may be used to reduce this error. Tablet-specific sample cells allow consistent positioning of tablets and reduce this error. Repeating measurements and using the mean (or median) spectrum for each tablet significantly reduces the spectral variability. The median calculation results in less weighting by odd spectra than does the mean.

Tablet spectra must usually be corrected for baseline shifting prior to analysis. Many techniques have been attempted, but second-derivative and multiplicative scatter correction calculations are most common. This baseline correction is critical even if an average tablet spectrum is used.

Curved surfaces, embossing, and scoring affect the spectrum of a tablet as positioning is varied, but the effect of such factors can be reduced by the methods just discussed. Natural variations in tablet mass and hardness also affect a tablet's spectrum through baseline shifting.

Work by Baxter involved a method of normalizing weight variations [132]. Baxter concluded that NIR reflectance spectra are "in essence a picture of active per unit area" and do not allow detection of differences in tablet weight. Thus, reference assay values should be normalized for tablet weight, multiplying the HPLC reference value by the theoretical tablet weight and dividing by the actual tablet weight. Values predicted from this calibration must then be denormalized by multiplying the NIR-predicted value by the tablet weight, divided by the theoretical weight. Baxter observed a reduction in residual values from 2.17 to 1.57% for 228 tablets for which active concentrations were predicted.

5.6 Conclusions

Tremendous advances have been made recently in the use of near-infrared spectroscopy for the analysis of pharmaceutical dosage forms. Just 25 years ago, near-infrared spectroscopy was used in a way that offered relatively few advantages over other analytical methods for the analysis of dosage form drug content, requiring extractions with organic solvents prior to sample analysis. With advances in instrumentation, software, and sample handling, rapid characterization of intact dosage forms has become a reality. The pharmaceutical industry is beginning to develop near-infrared methods to monitor many phases of the manufacturing process, from the arrival of bulk raw material at the loading dock to the inspection of tablets for final release.

Myths about the "black box" nature of this method have been debunked. As those involved in analytical methods development, process control, and quality assurance acquire a more thorough understanding of near-infrared spectroscopy and its capabilities, pharmaceutical applications will become more widespread. Near-infrared instruments are becoming faster, smaller, and less expensive, increasing their potential for application as process monitors in many phases of the manufacturing process. Pharmaceutical manufacturers are under increasing pressure to validate their processes and to provide extensive documentation of ongoing validation activities. Near infrared has proven to be a rapid and rugged analytical method capable of continuous online process monitoring, making it a valuable method to couple with ongoing validation activities.

In many ways, near-infrared spectroscopy is an ideal method for pharmaceutical process control, particularly for the analysis of intact dosage forms. As production costs, including analytical expenses, continue to increase, the advantages of NIR spectroscopy will become more attractive. With near-infrared spectroscopy, the pharmaceutical industry will move one step closer to zero-defect quality control, making the costs associated with the method's development well spent.

References

1. K. A. Martin, Recent Advances in Near-Infrared Reflectance Spectroscopy, *Appl. Spectrosc. Rev.*, 27, 325 (1992).
2. J. K. Drennen, E. G. Kraemer, and R. A. Lodder, Advances and Perspectives in Near-Infrared Spectrophotometry, *Crit. Rev. Anal. Chem.*, 22, 443 (1991).
3. E. W. Ciurczak, Uses of Near-Infrared Spectroscopy in Pharmaceutical Analysis, *Appl. Spectrosc. Rev.*, 23, 147 (1987).
4. M. Forina, M. C. Casolino, and C. de la Pezuela Martinez, Multivariate Calibration: Applications to Pharmaceutical Analysis, *J. Pharm. Biomed. Anal.*, 18, 21 (1998).
5. G. Reich, Near-Infrared Spectroscopy and Imaging: Basic Principles and Pharmaceutical Applications, *Adv. Drug Deliv. Rev.*, 57, 1109 (2005).
6. Y. Roggo, P. Chalus, L. Maurer, C. Lema-Martinez, A. Edmond, and N. Jent, A Review of Near Infrared Spectroscopy and Chemometrics in Pharmaceutical Technologies, *J. Pharm. Biomed. Anal.*, 44, 683 (2007).
7. J. Luypaert, D. L. Massart, and Y. Vander Heyden, Near-Infrared Spectroscopy Applications in Pharmaceutical Analysis, *Talanta*, 72, 865 (2007).

8. C. Gendrina, Y. Roggoa, and C. Collet, Pharmaceutical Applications of Vibrational Chemical Imaging and Chemometrics: A Review, *J. Pharm. Biomed. Anal.*, 48, 533 (2008).

9. J. Ali, K. Pramod, and S. H. Ansari, Near-Infrared Spectroscopy for Nondestructive Evaluation of Tablets, *Syst. Rev. Pharm.*, 1, 17 (2010).

10. A. C. Moffat, S. Assi, and R. A. Watt, Identifying Counterfeit Medicines Using Near Infrared Spectroscopy, *J. Near Infrared Spectrosc.*, 18, 1 (2010).

11. B. G. Osborne and T. Fearn, *Near-Infrared Spectroscopy in Food Analysis*, Longman Scientific and Technical, New York, 1986.

12. G. Patonay, ed., Dyes Used in NIR Bioanalysis, *Advances in Near-Infrared Measurements*, JAI Press, Greenwich, CT, 1993.

13. E. W. Ciurczak and R. Raghavachari, Applications of NIRS in Biopharmaceutical Analysis, in *Near-Infrared Applications in Biotechnology*, CRC Press, Boca Raton, FL, 2000.

14. W. R. Hruschka, Data Analysis: Wavelength Selection Methods, in *Near-Infrared Technology in the Agricultural and Food Industries*, ed. P. Williams and K. Norris, American Association of Cereal Chemists, St. Paul, MN, 2001.

15. D. A. Burns and E. W. Ciurczak, Identification of Raw Materials by NIRA, *Handbook of Near-Infrared Analysis*, 3rd ed., CRC Press, Boca Raton, FL, 2007.

16. J. Workman and L. Weyer, Qualitative Near Infrared Reflectance Analysis Using Mahalanobis Distances, *Practical Guide to Interpretive Near-Infrared Spectroscopy*, CRC Press, Boca Raton, FL, 2007.

17. J. E. Sinsheimer and A. M. Keuhnelian, Near-Infrared Spectroscopy of Amine Salts, *J. Pharm. Sci.*, 55, 1240 (1966).

18. N. Oi and E. Inaba, Analyses of Drugs and Chemicals by Infrared Absorption Spectroscopy. 8. Determination of Allylisopropylacetureide and Phenacetin in Pharmaceutical Preparations by Near Infrared Absorption Spectroscopy, *Yakugaku Zasshi*, 87, 213 (1967).

19. J. E. Sinsheimer and N. M. Poswalk, Pharmaceutical Applications of the Near Infrared Determination of Water, *J. Pharm. Sci.*, 57, 2006 (1968).

20. J. J. Rose, T. Prusik, and J. Mardekian, Near Infrared Multi-Component Analysis of Parenteral Products Using the InfraAlyzer 400, *J. Parenter. Sci. Technol.*, 36(2), 71 (1982).

21. H. L. Mark and D. Tunnell, Qualitative Near-Infrared Reflectance Analysis Using Mahalanobis Distances, *Anal. Chem.*, 57(7), 1449 (1985).

22. E. W. Ciurczak, Use of NIRS to Follow the Uniformity of Mixing in a Pharmaceutical Blend, presented at 7th Annual Symposium on NIRA, Technicon, Tarrytown, NY, 1984.

23. E. W. Ciurczak, Dissolution Studies of Multi-Component Dosage Forms Utilizing NIRS, presented at 9th Annual Symposium on NIRA, Technicon, Tarrytown, NY, 1986.

24. E. W. Ciurczak, Identification and Quality Assessment of Incoming Raw Materials and Pharmaceutical Mixtures, presented at AAPS National Meeting, Las Vegas, November, 1990.

25. E. W. Ciurczak, Following the Progress of a Pharmaceutical Mixing Study via Near-Infrared Spectroscopy, *Pharm. Technol.*, 15(9), 141 (1991).

26. E. W. Ciurczak, presented at Eastern Annual Symposium, Somerset, NY, November 1990.

27. E. W. Ciurczak, The Use of NIRS to Follow the Uniformity of Mixing in a Pharmaceutical Blend, presented at AAPS National Meeting, Las Vegas, November 1990.

28. *GMP Drug Report*, December 1999.

29. O. Y. Rodionova, Y. V. Sokovikov, and A. L. Pomerantsev, Quality Control of Packed Raw Materials in Pharmaceutical Industry, *Anal. Chim. Acta*, 6, 222 (2009).

30. G. Chen, F. Kamal, and J. Phillips, NIR Transmission Analysis of Solid Dosage Forms, *Am. Pharm. Rev.*, 12, 54 (2009).

31. Z. Q. Wen, G. Chen, Y. Luo, G. Li, P. Masatani, X. Cao, P. Bondarenko, and J. Phillips, Near-Infrared in the Pharmaceutical Industry, *Am. Pharm. Rev.*, 15, 26 (2012).

32. E. W. Ciurczak, Identification and Quality Assessment of Incoming Raw Materials and Pharmaceutical Mixtures, presented at AAPS National Meeting, Las Vegas, November 1990.

33. J. K. Drennen, Near-Infrared Spectroscopy: Applications in the Analysis of Tablets and Solid Pharmaceutical Dosage Forms, PhD thesis dissertation, University of Kentucky, 1991.

34. J. D. Kirsch and J. K. Drennen, Interactive Self-Modeling Mixture Analysis, *Appl. Spectrosc. Rev.*, 30, 139 (1995).

35. W. Windig and J. Guilment, Automated System for On-Line Monitoring of Powder Blending Processes Using Near-Infrared Spectroscopy. Part II: Qualitative Approaches to Blend Evaluation, *Anal. Chem.*, 63, 1425 (1991).

36. S. S. Sekulic, J. Wakeman, P. Doherty, and P. A. Hailey, A New PAT/QbD Approach for the Determination of Blend Homogeneity: Combination of On-Line NIRS Analysis with PC Scores Distance Analysis (PC-SDA), *J. Pharm. Biomed. Anal.*, 17, 1285 (1998).

37. T. Puchert, C. V. Holzhauer, J. C. Menezes, D. Lochmann, and G. Reich, NIR Analysis of Polymeric Packaging Materials, *Eur. J. Pharm. Biopharm.*, 78, 173 (2011).

38. F. C. Sanchez, J. Toft, B. Van den Bogaert, D. L. Massart, S. S. Dive, and P. Hailey, Chapter in Handbook of NIR Analysis, Monitoring Powder Blending by NIR Spectroscopy. Fresenius *J. Anal. Chem.*, 352, 771 (1995).

39. G. E. Ritchie, Qualitative Uses of NIRA in Pharmaceutical Analysis, presented at 9th International Diffuse Reflectance Conference, Wilson College, Chambersburg, PA, August 1998.

40. G. E. Ritchie, Near-Infrared Methods for the Identification of Tablets in Clinical Trial Supplies, presented at PittCon, New Orleans, March 2000.

41. M. A. Dempster, J. A. Jones, I. R. Last, B. F. MacDonald, and K. A. Prebble, Identification of Tablet Formulations inside Blister Packages by Near-Infrared Spectroscopy, *J. Pharm. Biomed. Anal.*, 11, 1087 (1993).

42. P. K. Aldridge, R. F. Mushinsky, M. M. Andino, and C. L. Evans, Comparison of Classification Approaches Applied to NIR-Spectra of Clinical Study Lots, *Appl. Spectrosc.*, 48, 1272 (1994).

43. A. Candolfi, W. Wu, D. L. Massart, and S. Heuerding, Implementation of a Simple Semi-Quantitative Near-Infrared Method for the Classification of Clinical Trial Tablets, *J. Pharm. Biomed. Anal.*, 16, 1329 (1998).

44. R. De Maesschalck and T. Van den Kerkhof, Identification of Actives in Multi-Component Pharmaceutical Dosage Forms via NIR Reflectance Analysis, *J. Pharm. Biomed. Anal.*, 37, 109 (2005).

45. E. W. Ciurczak and T. A. Maldacker, Determination of Enantiomeric Purity of Valine by Near Infrared Analysis, *Spectroscopy*, 1(1), 36 (1986).

46. D. E. Honigs, A New Method for Obtaining Individual Component Spectra from Those of Complex Mixtures, PhD dissertation, Indiana University, 1984.

47. D. E. Honigs, G. M. Heiftje, and T. Hirshfeld, Near Infrared Reflectance Analysis of Pharmaceutical Products, *Appl. Spectrosc.*, 38, 317 (1984).

48. R. G. Whitfield, The Use of Near Infrared Spectroscopy to Detect Counterfeit Medicines, *Pharm. Manuf.*, 3(4), 31 (1986).

49. T. Moffat, R. Watt, and S. Assi, NIR-Based Approach to Counterfeit-Drug Detection, *Spectrosc. Eur.*, 22, 6 (2010).

50. O. Y. Rodionova and A. L. Pomerantsev, Near Infrared Spectrophotometric Comparison of Authentic and Suspect Pharmaceuticals, *Trends Anal. Chem.*, 29, 795 (2010).

51. J. E. Polli, S. W. Hoag, and S. Flank, Near-Infrared Spectroscopy (NIRS) and Chemometric Analysis of Malaysian and UK Paracetamol Tablets: A Spectral Database Study, *Pharm. Technol.*, 46 (2009).

52. M. M. Said, S. Gibbons, A. C. Moffat, and M. Zloh, A Novel Sample Selection Strategy by Near-Infrared Spectroscopy-Based High Throughput Tablet Tester for Content Uniformity in Early-Phase Pharmaceutical Product Development. Note missing reference: J Pharm Sci. 2012 Jul;101(7):2502-11. doi: 10.1002/jps.23153. *Int. J. Pharm.*, 415, 102 (2011).

53. Z. Shi, J. G. Hermiller, T. Z. Gunter, X. Zhang, and D. E. Reed, Determination of Particle Size of Pharmaceutical Raw Materials Using Near-Infrared Reflectance Spectroscopy, *J. Pharm. Sci.*, in press, 2012.

54. T. Shintani-Young and E. W. Ciurczak, Use of NIRA in Polymeric Packaging Components, presented at PittCon, New Orleans, March 1985.

55. C. Kradjel and L. McDermott, Determination of the Thickness of Plastic Sheets Used in Blister Packaging by Near Infrared Spectroscopy: Development and Validation of the Method, NIR Analysis of Polymers, in *Handbook of Near-Infrared Analysis*, 3rd ed., CRC Press, Boca Raton, FL, 2007.

56. M. Laasonen, T. Harmia-Pulkkinen, C. Simard, M. Räsänen, and H. Vuorela, Relationship between Acute Toxicity in Mice and Polymorphic Forms of Polyene Antibiotics, *Eur. J. Pharm. Sci.*, 21, 493 (2004).

57. G. Ghielmetti, T. Bruzzese, C. Bianchi, and F. Recusani, *J. Pharm. Sci.*, 65(6), 905 (1976).

58. E. W. Ciurczak, Quantitative Determination of Polymorphic Forms in a Formulation Matrix Using the Near Infra-Red Reflectance Analysis Technique, presented at FACSS, Philadelphia, October 1985.

59. R. Gimet and T. Luong, *J. Pharm. Biomed. Anal.*, 5, 205 (1987).

60. C. E. Miller and D. E. Honigs, Discrimination of Different Crystalline Phases Using Near-infrared Diffuse Reflectance Spectroscopy, *Spectroscopy*, 4, 44 (1989).

61. E. Dreassi, G. Ceramelli, P. Corti, S. Lonardi, and P. L. Perruccio, Application of Near-Infrared Reflectance Analysis to the Integrated Control of Antibiotic Tablet Production, *Analyst*, 120, 1005 (1995).

62. P. K. Aldridge, C. L. Evans, H. W. Ward, S. T. Colgan, N. Boyer, and P. J. Gemperline, Near-IR Detection of Polymorphism and Process-Related Substances, *Anal. Chem.*, 68(6), 997 (1996).

63. K. De Braekeleer, F. Cuesta Sanchez, P. A. Hailey, D. C. A. Sharp, A. J. Pettman, and D. L. Massart, Influence and Correction of Temperature Perturbations on NIR Spectra during the Monitoring of a Polymorph Conversion Process prior to Self-Modelling Mixture Analysis, *J. Pharm. Biomed. Anal.*, 17, 141 (1998).

64. M. Savolainen, A. Heinz, C. Strachan, K. C. Gordon, J. Yliruusi, T. Rades, and N. Sandler, Screening for Differences in the Amorphous State of Indomethacin Using Multivariate Visualization, *J. Pharm. Pharmacol.*, 59, 161 (2007).

65. M. Otsuka, F. Kato, and Y. Matsuda, Determination of Indomethacin Polymorphic Contents by Chemometric Near-Infrared Spectroscopy and Conventional Powder X-Ray Diffractometry, *Analyst*, 126, 1578 (2001).

66. J. J. Seyer and P. E. Luner, Determination of Indomethacin Crystallinity in the Presence of Excipients Using Diffuse Reflectance Near-Infrared Spectroscopy, *Pharm. Dev. Technol.*, 64, 573 (2001).

67. M. Otsuka, F. Kato, Y. Matsuda, and Y. Ozaki, Comparative Determination of Polymorphs of Indomethacin in Powders and Tablets by Chemometrical Near-Infrared Spectroscopy and X-Ray Powder Diffractometry, *AAPS PharmSciTech*, 4, 1 (2003).

68. M. Otsuka, F. Kato, and Y. Matsuda, Quantifying Ternary Mixtures of Different Solid-State Forms of Indomethacin by Raman and Near-Infrared Spectroscopy, *AAPS PharmSciTech*, 2, 1 (2000).

69. A. Heinz, M. Savolainen, T. Rades, and C. J. Strachan, Comparative Evaluation of the Degree of Indomethacin Crystallinity by Chemoinfometrical Fourier-Transformed Near-Infrared Spectroscopy and Conventional Powder X-Ray Diffractometry, *Eur. J. Pharm. Sci.*, 32, 182 (2007).

70. I. Fix and K.-J. Steffens, Quantifying Low Amorphous or Crystalline Amounts of Alpha-Lactose-Monohydrate Using X-Ray Powder Diffraction, Near-Infrared Spectroscopy, and Differential Scanning Calorimetry, *Drug Dev. Ind. Pharm.*, 30, 513 (2004).

71. E. Räsänen, J. Rantanen, A. Jørgensen, M. Karjalainen, T. Paakkari, and J. Yliruusi, Novel Identification of Pseudopolymorphic Changes of Theophylline during Wet Granulation Using Near Infrared Spectroscopy, *J. Pharm. Sci.*, 90, 389 (2001).

72. T. D. Davis, G. E. Peck, J. G. Stowell, K. R. Morris, and S. R. Byrn, Modeling and Monitoring of Polymorphic Transformations during the Drying Phase of Wet Granulation, *Pharm. Res.*, 21, 860 (2004).

73. N. Sandler, J. Rantanen, J. Heinämäki, M. Römer, M. Marvola, and J. Yliruusi, Pellet Manufacturing by Extrusion-Spheronization Using Process Analytical Technology, *AAPS PharmSciTech*, 6, E174 (2005).

74. A. L. Kellya, T. Gougha, R. S. Dhumala, S. A. Halseyb, and A. Paradkara, Monitoring Ibuprofen-Nicotinamide Cocrystal Formation during Solvent Free Continuous Cocrystallization (SFCC) Using Near Infrared Spectroscopy as a PAT Tool, *Int. J. Pharm.*, 426, 15 (2012).

75. E. W. Ciurczak, Analysis of Solid and Liquid Dosage Forms via NIRS, presented at FACSS, Philadelphia, October 1985.

76. E. W. Ciurczak, Investigation of Particle Size Effects on NIRA, presented at 26th Annual Conference on Pharmceutical Analysis, Merrimac, Wisconsin, 1986.

77. B. R. Buchanan, E. W. Ciurczak, A. Grunke, and D. E. Honigs, Determination of the Enantiomeric Purity of Valine Using Near-Infrared Spectroscopy, *Spectroscopy*, 3(9), 54 (1988).

78. D. M. Mustillo and E. W. Ciurczak, Determination of Enantiomeric Purity of Valine by Near Infrared Analysis, presented at Eastern Analytical Symposium, Somerset, NY, November 1990.

79. E. W. Ciurczak and T. A. Dickinson, Further Parameters of a NIR/HPLC Detector. Part II: Non-Aqueous Solvents for Normal Phase and Size Exclusion Chromatography, *Spectroscopy*, 6(7), 36 (1991).

80. C. Kradjel and E. W. Ciurczak, Further Parameters of a NIR/HPLC Detector. Part I: Limits of Analytical Sensitivity for Reverse Phase Chromatographic Systems, presented at 1st Pan American Chemical Conference, San Juan, PR, October 1985.

81. P. Kubelka, New Contributions to the Optics of Intensely Light-Scattering Materials. Part II: Nonhomogeneous Layers, *J. Opt. Soc. Am.*, 38, 448 (1948).

82. G. Kortum, Effects of Particle Size on Scattering of Light, *Reflectance Spectroscopy: Principles, Methods, Applications*, Springer-Verlag, New York, 1969.

83. E. W. Ciurczak, presented at 1st Pan American Chemical Conference, San Juan, PR, October 1985.

84. E. W. Ciurczak, R. P. Torlini, and M. P. Demkowitz, Determination of Particle Size of Pharmaceutical Raw Materials Using Near-Infrared Reflectance Spectroscopy, *Spectroscopy*, 1(7), 36 (1986).

85. J. L. Ilari, H. Martens, and T. Isaksson, Determination of Particle Size in Powders by Scattering Correction in Diffuse Near Infrared Reflectance, *Appl. Spectrosc.*, 45(5), 722 (1986).

86. A. J. O'Neil, R. D. Jee, and A. C. Moffat, Measurement of the Cumulative Particle Size Distribution of Microcrystalline Cellulose Using Near Infrared Reflectance Spectroscopy, *Analyst*, 124, 33 (1999).

87. J. Rantanen and J. Yliruusi, Determination of Particle Size in a Fluidized Bed Granulator with a Near Infrared Set-Up, *Pharm. Pharmacol. Commun.*, 4, 73 (1998).

88. F. J. S. Nieuwmeyer, M. Damen, A. Gerich, F. Rusmini, K. van der Voort Maarschalk, and H. Vromans, Granule Characterization during Fluid Bed Drying by Development of a Near Infrared Method to Determine Water Content and Median Granule Size, *Pharm. Res.*, 24(10), 1854 (2007).

89. M. Otsuka, Chemoinformetrical Evaluation of Granule and Tablet Properties of Pharmaceutical Preparations by Near-Infrared Spectroscopy, *Chemometrics Intell. Lab. Syst.*, 82, 109 (2006).

90. M. W. J. Derksen, P. J. M. van de Oetelaar, and F. A. Maris, The Use of Near-Infrared Spectroscopy in the Efficient Prediction of a Specification for the Residual Moisture Content of a Freeze-Dried Product, *J. Pharm. Biomed. Anal.*, 17, 473 (1998).

91. M. S. Kamet, P. P. DeLuca, and R. A. Lodder, Near-Infrared Spectroscopic Determination of Residual Moisture in Lyophilized Sucrose through Intact Glass Vials, *Pharm. Res.*, 6(11), 961 (1989).

92. M. Brülls, S. Folestad, A. Sparén, and A. Rasmuson, In-Situ Near-Infrared Spectroscopy Monitoring of the Lyophilization Process, *Pharm. Res.*, 20, 494 (2003).

93. K.-I. Izutsu, Y. Fujimaki, A. Kuwabara, and N. Aoyagi, Effect of Counterions on the Physical Properties of l-Arginine in Frozen Solutions and Freeze-Dried Solids, *Int. J. Pharm.*, 301, 161 (2005).

94. W. Cao, C. Mao, W. Chen, H. Lin, S. Krishnan, and N. Cauchon, Differentiation and Quantitative Determination of Surface and Hydrate Water in Lyophilized Mannitol Using NIR Spectroscopy, *J. Pharm. Sci.*, 95, 2077 (2006).

95. R. J. Warren, J. E. Zarembo, C. W. Chong, and M. J. Robinson, Determination of Trace Amounts of Water in Glycerides by Near-Infrared Spectroscopy, *J. Pharm. Sci.*, 59(1), 29 (1970).

96. R. P. Torlini and E. W. Ciurczak, presented at PittCon, Atlantic City, March 1987.

97. G. X. Zhou, Z. Ge, J. Dorwart, B. Izzo, J. Kukura, G. Bicker, and J. Wyvratt, Determination and Differentiation of Surface and Bond Water in Drug Substances by Near Infrared Spectroscopy, *J. Pharm. Sci.*, 92(5), 1058 (2003).

98. E. W. Ciurczak and J. K. Drennen, Use of Chemometrics to Determine Surface and Bound Water by NIRS, *Spectroscopy*, 7(6), 12 (1992).

99. J. K. Drennen and R. A. Lodder, Determination of Hardness via NIRS, in *Advances in Near-Infrared Measurements*, ed. G. Patonay, JAI Press, Greenwich, CT, 1993, p. 93.

100. J. D. Kirsch and J. K. Drennen, Determination of Film Coated Tablet Parameters by Near Infrared Spectroscopy, *J. Pharm. Biomed. Anal.*, 13, 1273 (1995).

101. J. D. Kirsch and J. K. Drennen, Near-Infrared Spectroscopy: Applications in the Analysis of Tablets and Solid Pharmaceutical Dosage Forms, *Appl. Spectrosc. Rev.*, 30, 139 (1995).

102. W. R. Hruschka, Data Analysis: Wavelength Selection Methods, in *Near-Infrared Technology in the Agricultural and Food Industries*, ed. P. Williams and K. Norris, American Association of Cereal Chemists, 2001.

103. J. D. Kirsch and J. K. Drennen, presented at AAPS National Meeting, Seattle, WA, October 1996, paper APQ 1177.

104. J. D. Kirsch and J. K. Drennen, Nondestructive Tablet Hardness Testing by Near-Infrared Spectroscopy: A New and Robust Spectral Best-Fit Algorithm, *J. Pharm. Biomed. Anal.*, 19, 351 (1999).

105. K. M. Morisseau and C. T. Rhodes, Near-Infrared Spectroscopy as a Nondestructive Alternative to Conventional Tablet Hardness Testing, *Pharm. Res.*, 14(1), 108 (1997).

106. N. K. Ebube, S. S. Thosar, R. A. Roberts, M. S. Kemper, R. Rubinovitz, D. L. Martin, G. E. Reier, T. A. Wheatley, and A. J. Shukla, Application of Near-Infrared Spectroscopy for Nondestructive Analysis of Avicel Powders and Tablets, *Pharm. Dev. Technol.*, 4, 19 (1999).

107. Y. Chen, S. S. Thosar, R. A. Forbess, M. S. Kemper, R. L. Rubinovitz, and A. J. Shukla, Prediction of Drug Content and Hardness of Intact Tablets Using Artificial Neural Network and Near-Infrared Spectroscopy, *Drug Dev. Ind. Pharm.*, 27, 623 (2001).

108. M. Donoso, D. O. Kildsig, and E. S. Ghaly, Prediction of Tablet Hardness and Porosity Using Near-Infrared Diffuse Reflectance Spectroscopy as a Nondestructive Method, *Pharm. Dev. Technol.*, 8, 357 (2003).

109. M. Blanco, M. Alcalá, J. M. González, and E. Torras, A Process Analytical Technology Approach Based on Near Infrared Spectroscopy: Tablet Hardness, Content Uniformity, and Dissolution Test Measurements of Intact Tablets, *J. Pharm. Sci.*, 95, 2137 (2006).

110. M. Blanco and M. Alcalá, Content Uniformity and Tablet Hardness Testing of Intact Pharmaceutical Tablets by Near Infrared Spectroscopy. A contribution to Process Analytical Technologies, *Anal. Chim. Acta*, 557, 353 (2006).

111. M. Otsuka and I. Yamane, Prediction of Tablet Hardness Based on Near Infrared Spectra of Raw Mixed Powders by Chemometrics, *J. Pharm. Sci.*, 95, 1425 (2006).

112. H. Tanabe, K. Otsuka, and M. Otsuka, Theoretical Analysis of Tablet Hardness Prediction Using Chemoinformetric Near-Infrared Spectroscopy, *Anal. Sci.*, 23, 857 (2007).

113. C. D. Ellison, B. J. Ennisb, M. L. Hamada, and R. C. Lyon, Measuring the Distribution of Density and Tabletting Force in Pharmaceutical Tablets by Chemical Imaging, *J. Pharm. Biomed. Anal.*, 48, 1 (2008).

114. B. Igne, C. A. Anderson, and J. K. Drennen, Radial Tensile Strength Prediction of Relaxing and Relaxed Compacts by Near-Infrared Chemical Imaging, *Int. J. Pharm.*, 418, 297 (2011).

115. S. Sherken, Rapid Near-Infrared Spectrophotometric Method for Determination of Meprobamate in Meprobamate Tablets, *J. Assoc. Offic. Anal. Chem.*, 51, 616 (1968).

116. L. Allen, Quantitative Determination of Carisoprodol, Phenacetin, and Caffeine in Tablets by Near-IR Spectrometry and Their Identification by TLC, *J. Pharm. Sci.*, 63, 912 (1974).

117. A. F. Zappala and A. Post, Rapid Near IR Spectrophotometric Determination of Meprobamate in Pharmaceutical Preparations, *J. Pharm. Sci.*, 292 (1977).

118. P. Corti, E. Dreassi, G. Corbini, S. Lonardi, and S. Gravina, Application of Near Infrared Reflectance Spectroscopy to Pharmaceutical Control. I. Preliminary Investigation of the Uniformity of Tablets Content, *Analysis*, 18, 112 (1990).

119. J. K. Becconsall, J. Percy, and R. F. Reid, Quantitative Photoacoustic Spectroscopy of Propranolol/Magnesium Carbonate Powder Mixtures in the Ultraviolet and the Near-Infrared Regions, *Anal. Chem.*, 53, 2037 (1981).

120. E. W. Ciurczak and R. P. Torlini, Determination of Citamine in Pharmaceutical Granulations, *Spectroscopy*, 2(3), 41 (1987).

121. J. C. Chasseur, Analysis of Solid and Liquid Dosage Forms via NIRS, *Chim. Oggi*, 6, 21 (1987).

122. B. G. Osborne, Determination of Nicotinamide in Pre-Mixes by Near-Infrared Reflectance Spectrometry, *Analyst*, 112, 313 (1987).

123. R. A. Lodder and G. M. Hieftje, Subsurface Image Reconstruction by Near-Infrared Reflectance Analysis, *Appl. Spectrosc.*, 42, 1351 (1988).

124. P. Corti, E. Dreassi, C. Murratzu, G. Corbini, L. Ballerini, and S. Gravina, Application of NIRS to the Quality Control of Pharmaceuticals. Ketoprofen Assay in Different Pharmaceutical Formulations, *Pharm. Acta Helv.*, 64, 140 (1989).

125. P. Corti, E. Dreassi, G. Corbini, S. Lonardi, R. Viviani, L. Mosconi, and M. Bernuzzi, Application of Reflectance NIRS Spectroscopy to Pharmaceutical Quality-Control of Solid Binary Mixtures, *Pharm. Acta Helv.*, 65, 28 (1990).
126. J. A. Ryan, S. V. Compton, M. A. Brooks, and D. A. C. Compton, Rapid Verification of Identity and Content of Drug Formulations Using Mid-Infrared Spectroscopy, *J. Pharm. Biomed. Anal.*, 9, 303 (1991).
127. P. Corti, L. Savini, E. Dreassi, S. Petriconi, R. Genga, L. Montecchi, and S. Lonardi, Application of NIRA to Verification of Tablets for Clinical Trials, *Process Control Qual.*, 2, 131 (1992).
128. M. Blanco, J. Coello, H. Iturriaga, S. Maspoch, and C. De La Pezuela, Determination of Ascorbic Acid in Pharmaceutical Preparations by Near Infrared Reflectance Spectroscopy, *Talanta*, 40, 1671 (1993).
129. N.-H. Duong, P. Arratia, F. Muzzio, A. Lange, J. Timmermans, and S. Reynolds, A Homogeneity Study Using NIR Spectroscopy: Tracking Magnesium Stearate in Bohle Bin-Blender, *Drug Dev. Ind. Pharm.*, 29, 679 (2003).
130. M. Ito, T. Suzuki, S. Yada, H. Nakagamib, H. Teramotoc, E. Yonemochia, and K. Terada, Development of a Method for Nondestructive NIR Transmittance spectroscopic Analysis of Acetaminophen and Caffeine Anhydrate in Intact Bilayer Tablets, *J. Pharm. Biomed. Anal.*, 53, 396 (2010).
131. M. Boiret, L. Meunier, and Y.-M. Ginot, Tablet Potency of Tianeptine in Coated Tablets by Near Infrared Spectroscopy: Model Optimisation, Calibration Transfer and Confidence Intervals, *J. Pharm. Biomed. Anal.*, 54, 510 (2011).
132. M. Baxter, Determination of Active Ingredient in the Film Coating of a Tablet by NIRS, presented at Eastern Analytical Symposium, Somerset, NJ, 1994, paper 353.

chapter 6

Validation issues

6.1 Introduction

The major portion of this book discusses the theory, mathematics, applications, and mechanics of near-infrared (NIR) analysis. However, the most cleverly designed NIR method is of limited use to the pharmaceutical world unless it has been validated and accepted by the agency in charge in situations where NIR has been determined as mitigating a risk associated with the manufacturing process. In the United States, this agency is currently the Food and Drug Administration (FDA). Numerous regulatory documents and a wide variety of books [1–9], papers, and technical standards provide guidelines for method validation. In short, assuming that an appropriate quality assurance program is in place, three criteria must be met: validation of the software, validation of the hardware, and then, and only then, validation of the NIR spectroscopic method (or any analytical method).

The U.S. Food and Drug Administration regulates the manufacture and distribution of pharmaceuticals, foods, cosmetics, medical devices, and biological products under the authority of the Food, Drug, and Cosmetic Act (21 USC) and the Public Health Service Act (42 USC). General requirements for validation can be found in Title 21 of the Code of Federal Regulations (CFR) under the headings "Good Manufacturing Practices (GMPs)," "Good Laboratory Practices (GLPs)," and "Good Clinical Practices (GCPs)."

The United States Pharmacopeia (USP) [10], in a chapter on validation of compendial methods, defines "analytical performance parameters" (accuracy, precision, specificity, limit of detection, limit of quantitation, linearity and range, ruggedness, and robustness) that are to be used for validating analytical methods. A USP general chapter on near-infrared spectrophotometry [11] addresses the suitability of instrumentation for use in a particular method through a discussion of operational qualifications and performance verifications.

The American Society for Testing and Materials (ASTM) provides considerable direction regarding appropriate methodology for establishing spectrophotometer performance tests and multivariate calibration [12–14]. In addition, the International Union of Pure and Applied Chemistry,

Analytical Chemistry Division, published several technical reports discussing method development and validation [15–17].

In 1996 and 1997 the International Conference on Harmonization (ICH) published guidelines for analytical method validation [18]. These documents present a discussion of the characteristics for consideration during the validation of the analytical procedures included as part of registration applications submitted within the European Union, Japan, and the United States. Regulatory agencies have emitted documents, largely inspired by ICH documents. It is the case of the FDA [19, 20], the European Medicines Agency (EMA) [21, 24], and the Australian Pesticides and Veterinary Medicines Authority [25], to name just a few.

Regardless of the numerous sources providing general direction to the method developer, *validation* is an extremely complex regulatory issue that is not easily defined. Historically, a NIR spectroscopist considered validation to be the process of verifying the correlation between NIR results and the compendial or reference method using appropriate statistical or chemometrics routines. The term *validation*, in the regulatory context of a NIR method, refers to the establishment of appropriate data and documentation to certify that the method performs as intended. This certainly encompasses the more specific practice of proving that a NIR method is providing results that correlate with reference methods.

6.1.1   Quality assurance

The first issue for any pharmaceutical analyst considering method validation is the general concern regarding a quality assurance (QA) program. A formal quality assurance program is mandatory in today's highly regulated pharmaceutical industry. Development of a valid quality assurance program requires involvement of laboratory personnel at all levels. Management must be committed to such an undertaking and establish a mechanism for its development and implementation.

The developers of a quality assurance program must articulate the goals, document the quality assurance procedures that are already in place, and establish the new QA program. The quality assurance program for a laboratory must be maintained in a comprehensive manual that will typically contain current good laboratory practices (cGLPs), current good manufacturing practices (cGMPs), standard operating procedures (SOPs), and guidelines for implementing the program within the organization's unique structure. The FDA's "Good Laboratory Practices for Nonclinical Laboratory Studies" (21 CFR Part 58) assure that such studies are conducted according to scientifically sound protocols and proper procedures. Requirements pertaining to such laboratories where nonclinical studies are carried out include the following:

- Inspections for compliance must be permitted.
- All personnel must have adequate education, training, and experience.
- Resumes, job descriptions, and training records must be on file for all personnel; sufficient staff must be available to conduct studies.
- Management must designate a study director, who establishes a quality assurance unit and is responsible for validating results.
- The study director, who must have appropriate training and experience, has overall responsibility for conduction of study, assures that protocols are followed, and is solely responsible for authorizing changes in protocols.
- The quality assurance unit is mandatory and separate from study management.
- Facilities must be suitable for measurements and handling of test items; administrative and personnel facilities must be adequate.
- A written protocol is required for each study that defines all aspects of how it is to be conducted and the records that are to be kept, the study must be conducted according to the protocol and monitored for conformance, and any changes must be explainable.

6.1.2   Qualification and verification of NIR instruments

An important part of any validation protocol for NIR or other instrumental methods is the process of instrument qualification. In any GMP/GLP facility, instrument performance must be validated. This instrument performance validation, also called instrument qualification, involves three phases [26]: installation qualification (IQ), operational qualification (OQ), and performance qualification (PQ).

Installation qualification is documented verification that all key aspects of software and hardware installation adhere to appropriate codes and approved design intentions and that the recommendations of the manufacturer have been suitably considered. This involves documenting that the system has been installed as specified. Operational qualification is documented verification that the equipment or system operates as intended throughout representative or anticipated operating ranges. In practice, this is ensuring that the installed system works as specified. Sufficient documentary evidence must exist to demonstrate this. Performance qualification is documented verification that the processes and the total process-related system perform as intended throughout all anticipated operating ranges.

In practice, this is ensuring that the system in its normal operating environment produces acceptable product and sufficient documentary evidence exists to demonstrate this. This is typically performed on a complete system. Instrument vendors should usually perform IQ and OQ testing during installation of a new instrument. Performance qualification testing is more often performed by laboratory personnel through documenting the accuracy, linearity, and precision of the system under typical operating conditions. After installation and qualification, the instrument requires regular maintenance and calibration. The instrument should be maintained on a preventive maintenance plan and calibrated periodically according to a documented schedule.

The schedule for preventive maintenance and performance verification/ recalibration will depend on the instrument, its environment, and its function. The USP chapter on NIR spectrophotometry [11] suggests that "the purpose of instrument qualification is to assure that an instrument is suitable for its intended application and, when requalified periodically, it continues to function properly over extended time periods." Performance verification includes "a validation based on a quality of fit to an initial scan or group of scans included in the instrument qualification." With such testing, it is expected that reference standard spectra collected on a new or newly repaired, properly operating instrument should be identical to spectra of those standards collected over time, providing a basis for evaluating the long-term stability of a NIR instrument.

It is suggested that instrument qualification be performed every 6 months or following major repairs or optical reconfigurations. Qualification tests include wavelength accuracy, absorbance-scale linearity, and high- and low-light flux noise tests.

Performance verification is carried out on a system that is configured for actual measurements. Wavelength accuracy, absorbance-scale linearity, and high-light-level noise are to be tested monthly; wavelength accuracy is required and absorbance-scale linearity is optional prior to any data collection. Performance is verified through matching reference spectra to those spectra collected during previous instrument qualification tests. Wavelength accuracy is tested using an appropriate wavelength standard. The chapter recommends the SRM 1920, using peaks that occur at 1261, 1681, and 1935 nm for the traditional NIR region.

Table 6.1 displays recommended specifications for wavelength accuracy, photometric linearity, and spectrophotometric noise levels for pharmaceutical applications. The first step in validating any NIR method is to test the suitability of these specifications for a given application. Wavelength accuracy tests conducted using appropriate external standards will prevent potential problems that could occur with proprietary internal calibration protocols. The exact nature of any calibration standard must be noted in a validation protocol. Rare earth oxides and glass standards are candidates for such calibration.

Photometric linearity is tested using a set of standards with known relative transmittance or reflectance, depending on the application. Repeated measurements on stable standards give information about the

Table 6.1 Recommended NIR Instrument Specifications

Tolerances	±1 nm at 1200 nm
	±1 nm at 1600 nm
	±1.5 nm at 2000 nm
Noise	Measured for 100 nm segments from 1200–2200 nm
Average root mean square (RMS) noise level at high light flux	Less than 0.3×10^{-3}; no RMS where absorbance is greater than 0.8×10^{-3}
Average RMS noise level at low light flux	Less than 1×10^{-3}; no RMS where absorbance is greater than 2×10^{-3}
Photometric linearity	A_{obs} versus A_{ref} at 1200, 1600, and 2000 nm Slope = 1.0 ± 0.05, intercept = 0.0 ± 0.05

long-term stability of an instrument. For analyte systems with absorbance levels below 1.0, at least four reference standards are to be used in the range of 10% to 90%. When analytes provide absorbance values that exceed 1.0, the USP chapter recommends adding a 2% or 5% standard, or both.

Spectrophotometric noise tests include measuring spectra of high- and low-reflectance (or transmittance) reference materials. Peak-to-peak noise and root mean square (RMS) noise levels are acceptable parameters for evaluation. The preferable measurement involves tabulating the RMS noise in successive 100 nm spectral segments. Instrument noise is usually evaluated as a function of wavelength by using a reference standard as the sample and the background reference. According to the USP chapter, performance verification is performed with a single external performance verification standard, after operational qualification has shown that the equipment is acceptable for use. Such standards should match the format, size, and shape of actual samples as closely as possible. Frequent analysis of the standard allows reverification of the instrument performance on a continuing basis. Ideally, such performance verification is run before and after each series of analytical measurements to verify both wavelength accuracy and photometric accuracy.

American Society for Testing and Materials (ASTM) document E1866-97, entitled *Standard Guide for Establishing Spectrophotometer Performance Tests* [13], describes basic procedures that can be used to develop such performance tests. The document provides valuable insight into appropriate test conditions, samples for performance testing, and the various univariate and multivariate measures of spectrophotometer performance, some of which are discussed in the proposed USP chapter.

Performance test results should be plotted on charts for trend analysis. Statistical quality control charting methods are used to detect statistically significant changes in instrument performance. Action limits can be set for test charts based on historical data so that appropriate repairs can be made when necessary. Examples of unacceptable instrument performance may be valuable in setting action limits for future performance tests.

6.1.3   Method validation

With a quality assurance program in place and a qualified instrument ready for analysis, the method development lab is ready to develop, validate, and transfer analytical methods to the end user. McGonigle [27] discusses five essential principles to ensure successful method transfer: documentation, communication, acceptance criteria, implementation, and method modification and revalidation.

Documentation is the key to validation, and in fact, the development laboratory must provide the end user or designated laboratory with the following documentation:

- A written procedure
- A method validation report
- System suitability criteria

The procedure must provide adequate detail so that a designated laboratory can reproduce the method without deviation. The validation report includes the experimental design and the data that justify the conclusion that the analytical method performs as intended. Minimum acceptable performance criteria must also be defined in the report.

McGonigle discusses the significance of communication between the method development lab and end user at early stages of method development in light of the FDA's preapproval inspection process placing importance on verifying that the official method is consistent with that included in the new drug application. The designated laboratory is responsible for issuing and following SOPs that define their criteria for accepting an analytical method. The method transfer report includes data generated in accordance with those SOPs. The most important principle, according to McGonigle, is the implementation of the new analytical method as written and validated. Finally, McGonigle discusses the types of method modifications that require revalidation. Any modification that could potentially affect the accuracy, precision, linearity, and range of a method requires revalidation of these parameters.

6.2   International conference on harmonization

6.2.1   Validation guidelines

The International Conference on Harmonization (ICH) is attempting to standardize regulations and testing for pharmaceutical products. This conference was formed to bring about one set of rules and guidelines for the member states, enabling trade of pharmaceutical products between member nations to proceed more smoothly.

The ICH committee consists of six members: the European Commission of the European Union (EU), the European Federation of Pharmaceutical Industries and Associations (EFPIA), the Ministry of Health and Welfare, Japan (MHW), the Japan Pharmaceutical Manufacturers Association (JPMA), the U.S. Food and Drug Administration (FDA), and the Pharmaceutical Research and Manufacturers of America (PhRMA). Additional members include observers from the World Health Organization (WHO), European Free Trade Association, and Canada (Health Canada). The International Federation of Pharmaceutical Manufacturers and Associations (IFPMA) has been closely involved with ICH since its inception. This mixture of governmental agencies and professional representatives is performing an exceptional task: developing one set of guidelines for three major markets.

The publication of ICH guidelines [18] in 1996 and 1997 created a new paradigm for method development. Through the early 1990s, both academic and industrial workers published a large number of NIR applications. After 1997, however, the need to implement methods under ICH guidelines forced NIR spectroscopists to consider methods for validating their work in such a way as to gain approval in a worldwide market.

Using the ICH guidelines as a template to format a validation scheme for NIR spectroscopic methods is certainly a good idea. In assessing the appropriateness of using every specific parameter and criterion for accepting or rejecting NIR data, however, there are some inconsistencies. Namely, the guidelines are focused on unifying the validation of chromatographic methodologies, rather than NIR methodologies. While these parameters, per se, can be applied to many nonchromatographic methods, clearly some of the approaches and criteria cannot be applied to NIR methodologies. Some firms are using the NIR techniques for information only, i.e., in go/no-go tests and in-process testing. Since prerelease assays are not subject to the same validation guidelines as final release assays, these are the first to have been implemented. Major cost and time savings, however, will only be realized in final release testing. The FDA, EMA, and other agencies have published guidelines describing how process analytical technologies and, more particularly, near-infrared spectroscopy should be implemented [20–25].

A major concern of the agency has to do with the sensitivity of the NIR methods to process changes, including variation in raw materials and equipment. Inherent in validated chromatographic methods is the fact that changes in instrument response reflect only the concentration of the active ingredient, regardless of process or vendor changes to the raw materials. The reason that UV/Vis spectroscopy works so well and is accepted with full confidence by most workers in the field is that it is based on well-known and established laws of physics. The Beer-Bouger-Lambert law works.

However, in the case of NIR spectroscopy, and when it comes to non-linear relationships in complex, single-, or multicomponent mixtures, the physics are a little less understood (or understood by fewer workers). Certainly, the mathematical techniques used to unravel the mysteries of complex nonlinear instrument signals are even less widely understood.

What this means is that an educational process should be undertaken for industrial analysts and the agency addressing the complexities of NIR spectroscopy. Analysts must educate themselves about chemometrics before practicing NIR spectroscopy. They must also develop guidelines and create appropriate laws to regulate a field that is just beginning to be understood and appreciated.

Figure 6.1 presents what may be referred to as the near-infrared method validation pyramid. Ritchie and Ciurczak [28] presented this idea at the 9th International Diffuse Reflectance Conference (Chambersburg, Pennsylvania) in 1998. The figure shows that a NIR result, whether generated from a qualitative or quantitative method, can be traced back to a calibrated instrument and a calibrated computer hardware and software system. In addition, it reveals that the result was generated using a method that was found suitable for its intended use. Observed in this manner,

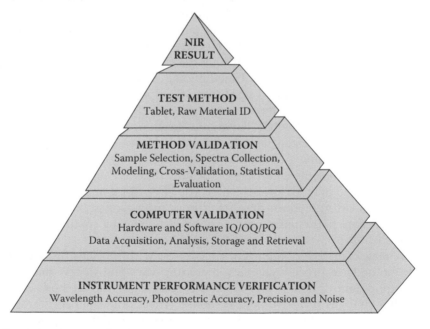

Figure 6.1 Diagram displaying the necessary steps in analytical method validation.

even a simple two-cluster NIR discriminatory method can be developed in a way that satisfies the current good manufacturing practices that apply to analytical methodologies used for pharmaceutical analysis.

6.2.2   ICH Recommended parameters for method validation

6.2.2.1   Specificity

Specificity is the ability to assess unequivocally the analyte in the presence of components that may be expected to be present. A major problem with utilizing NIR to accurately determine concentrations of active ingredients in solid dosage forms is that the analysis is performed without removing the analyte from its matrix. The very property that makes NIR so attractive, allowing analysis of intact dosage forms and intermediate products without dissolution/extraction of the active agent, is also the issue that could potentially limit its application to pharmaceutical analysis, if we are restricted to current validation regulations. The spectrum generated by the instrument represents all materials present in the formulation, including the active. Clearly then, specificity is a major validation hurdle that must be overcome if NIR is to be used as a release testing technique.

One mechanism that is proposed to ameliorate this problem is to first identify the product being tested before quantifying the analyte of interest. Using a well-established algorithm for spectral matching, such as principal component analysis followed by an appropriate qualitative pattern recognition routine like the Mahalanobis distance calculation, would be a viable approach.

In general, this process of product identification can be instituted by analyzing several lots of product with varying analyte values. The lots are identified by current analytical methods (usually high-performance liquid chromatography (HPLC)) and labeled "Product XYZ—acceptable range." If out-of-specification (OOS) lots are available (either synthetic or production lots), these are similarly labeled as high, low, etc. The appropriate qualitative algorithm is then used to identify and fingerprint the good product. This identification routine would be performed before any quantitative prediction, assuring a correct value.

6.2.2.2   Linearity

A second challenge to overcome is the linearity issue. In chromatographic and simple UV/Vis (Beer's law) methods, linearity of response to analyte is shown by making serial dilutions of a concentrated standard. These solutions are either injected into the HPLC or read on a spectrometer. The concentration is plotted against the area/peak height of the eluted peak (HPLC) or the absorbance value (UV/Vis). With NIR, things are not that simple.

Usually, the linearity of a NIR spectroscopic method is determined from the multiple correlation coefficient (R) of the NIR predicted values of either the calibration or validation set with respect to the HPLC reference values. It may be argued that this is an insufficient proof of linearity since linearity (in this example) is not an independent test of instrument signal response to the concentration of the analyte. The analyst is comparing information from two separate instrumental methods, and thus simple linearity correlation of NIR data through regression versus some primary method is largely inappropriate without other supporting statistics.

An analyst could extract a pure signal, directly correlated to the analyte of interest (matrix effect) only with major effort; it is illogical to compare the dosage form spectra to pure standards or solutions of standards at various concentrations. Calculating a correlation coefficient between two distinctly different methods is not the same as the linearity for chromatographic methods.

Poor values for linearity are usually the result when only production samples are used for development of quantitative calibrations. The low R values are a result of the narrow range of active concentrations due to the quality that is inherent in pharmaceutical processes. A typical pharmaceutical process gives samples that vary little from the nominal value. Typical ranges for tablet assays fall within a range of 97 to 102% of label claim. This 5% range, coupled with errors in the HPLC analysis, lead to a poor correlation line. In HPLC or UV/Vis spectroscopic methods, an R value approaching unity is common: the USP recommends a value no lower than 0.995. A typical NIR linearity, based solely on actual production samples, is often closer to 0.8 than 1.0 due primarily to the narrow range.

A potentially more appropriate statistical tool, the Durbin–Watson statistic [29], can be used to assess the linearity of the NIR quantitative method. This statistic allows the analyst to establish the lack of intercorrelation between data points in the regression. The correlation coefficient R only describes the *tendency* of the line, not the *trueness of fit* to a linear model. If there is no intercorrelation of the residuals described by the Durbin–Watson statistic, then a linear model is appropriate and may be used.

The Durbin–Watson statistic is more of a test for nonlinearity, calculated from residuals obtained from fitting a straight line. The statistic evaluates for sequential dependence in which error is correlated with those before and after the sequence. The formula is

$$d = \frac{\sum_{u=2}^{n} (e_u - e_{u-1})^2}{\sum_{u-1}^{n} e_u^2} \qquad (6.1)$$

where d is the Durbin–Watson statistic, e_u is the uth residual (estimate of error) between the reference (HPLC assay) and the predicted (NIR prediction), e_{u-1} is the $u - 1$ residual, and n is the number of samples in validation set.

It is known that:

- $0 \leq d \leq 4$ always.
- If successive residuals are positively serially correlated, that is, positively correlated in their sequence, d will be near zero.
- If successive residuals are negatively correlated, d will be near 4, so that $4 - d$ will be near zero.
- The distribution of d is symmetric about 2.

It may be that for products where out-of-specification samples may be manufactured (usually in development labs or pilot plants) to extend the range, the correlation coefficient will approach 1. And, in isolated cases, small ranges of sample values may yield a highly correlated series of values. Despite these cases, a large number of equations will generate poor R values while simultaneously giving excellent residuals. It is these cases, where poor correlation values are evident, that alternative methods must be used on a case-by-case basis insofar as validation for FDA is concerned. Perhaps the Durbin–Watson statistic is suitable for these cases.

6.2.2.3   Range

The ICH guidelines recommend a minimum range of 80 to 120% of the test concentrations for assay of a drug substance or a finished product and 70 to 130% of the test concentration for content uniformity. Therein lies what is one of the most significant issues prohibiting a uniform approach in implementing quantitative NIR techniques in the industry.

As mentioned in a previous section, the inherent quality of pharmaceutical processes leads to very narrow concentration ranges for production samples—the typical range covering only something in the neighborhood of 5%. The important question that we must ask, and ultimately find an answer for, is: How do we develop a quantitative calibration that meets ICH criteria (with an appropriately broad range of concentrations) when the production samples provide a range that is far short of what is required? The obvious answer appears to be that we might just manufacture samples with the necessary range of concentrations. In reality, the solution is not so simple. The first problem is that the method development team will probably find it impossible to convince management to provide out-of-specification product in the production environment. If that is a problem, then why can't we manufacture our out-of-specification product in the pilot plant or even in the laboratory?

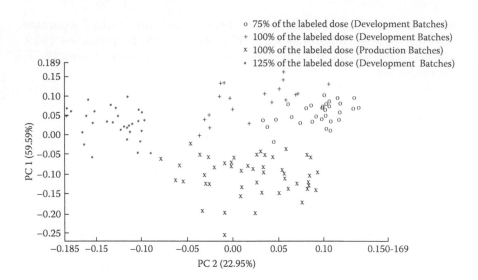

Figure 6.2 An example of the spectral shift arising from unique production lines is apparent in this principal component plot where significant displacement is observed between pilot and production batches of product containing identical drug concentrations. Each point represents the spectrum of a single tablet.

Now a second problem is evident. Regardless of the fact that identical raw materials (relative to the production samples) can be used in the lab or pilot plant, the process signature is often so different from laboratory to production to pilot scale that significant calibration errors will arise. For example, the dwell time, compression force, and feed rate variations that exist between the different scales of manufacturing can cause significant variations in sample spectra, even when formulations are identical. Thus, expansion of the range of concentration values through the use of laboratory or pilot plant samples is, in most cases, impractical. The NIR spectra of these samples will seldom represent the population eventually to be predicted.

Figure 6.2 displays an example of the possible spectral shift that can occur between production and pilot samples. It is possible that the range problem is the single most significant issue prohibiting a uniform approach in implementing quantitative NIR techniques in the industry. Until industrial and agency analysts address this issue, questions regarding the approval of quantitative NIR methods for release testing will remain.

6.2.2.4 *Ruggedness/robustness*

The robustness of any method may be measured by inflicting small, deliberate variations in the method and ascertaining whether any changes in the predictions gleaned are significant. When using, for instance, a tablet transmission module, it is assumed that the holder is unequivocal in its positioning of the tablet or capsule. It has been machined for the particular product, and few other tablets will fit within. (Since most capsules are sized as #0, #1, #2, etc., and are standard, the specificity tests will have to be run in conjunction with the analysis.) The only changes that might be imposed are turning the tablet over from one side to the other. It may be possible for a symmetric dose to turn it north to south as well.

Compliance with 11 CFR Part 35 should provide adequate proof of the ruggedness and robustness of the system software. Method developers must consider tests to evaluate algorithm stability during the development and validation processes.

6.2.2.5   Precision

The precision of an analytical procedure expresses the closeness of agreement between a series of measurements obtained from multiple sampling of the same homogenous sample under the prescribed conditions. Precision is considered at three different levels: repeatability, intermediate precision, and reproducibility.

Since this discussion is specific to intact dosage forms, repeatability takes on a slightly different color than for HPLC methods. While sample preparation is usually significantly reduced for NIR methods in comparison to traditional wet chemical methods, thus reducing method error from dissolution, extraction, and the like, the diffuse reflectance and diffuse transmittance methods common to NIR practice are susceptible to other factors that can ultimately reduce precision. The "repack error" once described for ground grain samples that were repacked into the traditional NIR sample cup may more aptly be described as a "positioning error" for today's nondestructive analyses of intact dosage forms. Appropriate sample handling and various mathematical routines allow minimization of the spectral variability and imprecision arising from this type of error.

6.2.2.6   Calibration transfer

Of similar significance to the range issue, suitable approaches for calibration transfer must be accepted by regulators for successful implementation of NIR spectroscopy on a large scale within the pharmaceutical industry. Regardless of instrument quality or the utmost discipline and care during method development, calibration transfer will be a reality if a calibration is used over any length of time.

While method transfer, discussed earlier in this chapter, refers to the process of implementing a newly developed method in a designated laboratory, the term *calibration transfer* is used in several contexts. One context, in fact, is similar to the usual definition of method transfer. A question that arises when NIR equations are concerned is: Can a

second instrument, using the same equation, give predictions comparable to those of the instrument on which the equation was developed?
Nonpharmaceutical applications (i.e., agriculture, textiles, process foods,
and tobacco) are routinely transferred between instruments. The only difference between those applications and the pharmaceutical applications
that the reader is concerned with is the paperwork involved with any FDA
application. The analyst must develop a protocol for successful calibration
transfer. The exact optimal protocol is currently open for debate.

Although today's manufacturers of NIR instruments do provide
instruments that are surprisingly consistent and precise from one device
to the next, the industrial method development chemist is likely to encoun-
ter problems when transferring methods from the laboratory to the production floor on a large scale. The instrumental differences that lead to
such problems may or may not be of adequate significance to be identified
during the instrument qualification tests. Appropriate use of reference
standards for performance qualification can certainly reduce the occur-
rence of such problems. However, in certain cases, one of the many calibration transfer algorithms that have been discussed in the literature over
the last decade may be required for successful transfer and implementation of a calibration.

Calibration transfer methods can also find wide application for what
is also known as calibration maintenance. It must be assumed that the
physical process, the instrument, the raw materials, or the production
personnel will change with time. Any implementation of a NIR method
without these safeguards is not only poor science, but will give NIR a bad
reputation when (not if) the equation fails to give a proper prediction.

It is widely acknowledged that there should be some mechanism for
checking the calibration equation after implementation. There are two
types of triggers that would cause an equation to be checked. The first is a
planned time point, similar to a scheduled calibration of a pump or detector for HPLC. In this case, for instance, at 6 months after implementation, a number of samples are scanned and predicted via NIR. These same
tablets are then subjected to HPLC analysis. The answers are compared
and statistical means are used to show that they are or are not within the
parameters of the initial installment. Should they not check out, a written procedure (calibration transfer protocol) should be in place for the
regeneration of the equation to account for any changes in the process or
analytical instrumentation. If there is no significant difference, then the
equation continues to be used until the next check period.

A second, and perhaps more important, safeguard is the use of control
charts for trend analysis. This plot will show any drift from the median
value (nominal value or actual average value) alerting whoever is responsible for calibration maintenance before the values actually drift out of
acceptance range. It is possible that the reason for the drift is the process

itself. In this case, a service has been performed for production. If tablets/ capsules (or any other product) are analyzed by the referee method and found to provide the expected value, the calibration transfer protocol may again be followed to correct the calibration.

Two landmark articles published by H. Mark et al. and G. E. Ritchie et al. [30, 31] demonstrate how ICH Q2 principles can be applied when developing a near-infrared method. The first part of the series describes the method validation concepts as described in this section and proposes additional validation protocols. The second part focuses on the implementation of assay and content uniformity methods for solid dosage forms. Authors present at length how each criterion was validated and how they meet ICH and FDA requirements. For instance, repeatability was tested with the predictions of 13 repeats of tablets at three different concentration levels (80, 100, and 120% w/w). In addition to the stated guidelines, authors used ASTM E1655-97 [12] to provide statistics analysis that is not directly provided in ICH Q2(R1).

Additional examples of method validation were published by M. Blanco et al. [32] in the implementation of a NIR model for active prediction across pharmaceutical processes or by Peinado et al. [33], who proposed a method validation approach for in-line measurements.

6.3 Historical perspective

Whitfield [34] was one of the first NIR spectroscopists to discuss the use of NIR diffuse reflection analysis for veterinary products in 1986. Even then, he recognized the need for specificity and included an identification step in the analysis. This discriminant analysis program, named DISCRIM (Technicon, Tarrytown, New York), was an early version of the types of algorithms available commercially now.

Three major advances have moved the technique closer to equal utility with chromatographic methods for release testing. The first advance was the development of suitable hardware. While useful for food and agricultural products, early interference-type filter instruments and scanning grating monochromators were neither sensitive nor quiet (instrument noise, not operating noise) enough for certain pharmaceutical applications. The development of stable, sensitive, and precise instruments has obviated the inherent problems with reflectance spectroscopy.

Extremely quiet and rapid monochromators are now available from numerous vendors. As covered in detail in Chapter 2, top-notch monochromators (grating and interferometer), diode arrays, accousto-optic tunable filters, and modern interference filter instruments now exist. Fiber optic probes, multiple detector modules, and transmission attachments all lend themselves to superior sample handling of powders and solid dosage forms.

The second advance was software. Prior to software packages supplied by Technicon (became Bran + Leubbe, now defunct) and Pacific-Scientific (FOSS-NIRSystems, now Metrohm NIRSystems), little was available for calculating analyte concentration. Extraction of analyte into solvents and reading the solution in a cuvette was all that existed until about 1980. Then, Karl Norris developed multivariate equations where several wavelengths were combined in what amounted to a multiterm Beer's law equation. Until the two mentioned vendors released their (nonvalidated) software packages, calculations were basically performed by hand. In addition, equations were not secured, validation trails were not made, and there was no security in terms of system access.

New software packages are available from all instrument vendors as well as many third-party vendors. Many provide the necessary software to document the development and validation of calibration methods. The companies are also often willing to allow the FDA to examine their source codes, another part of 21 CFR Part 11 regulations. Security exists for all major software packages: log-in codes, storage of spectra, and electronic trails for every assay. Most pharmaceutical companies are wrestling with software validation, with dozens of standard operating procedures dealing with the issue. Although the validation of NIR methods is currently a challenging issue due to a lack of appropriate regulatory guidelines, the problems will work themselves out in time as the methodology matures and as adequate numbers of analysts and regulators become more familiar with the technology.

The third advance was the publication of ICH guidelines for validating analytical methods used in the pharmaceutical analytical laboratory. This "menu" is roughly modeled after the steps used for validation of high-performance liquid chromatographic methods: linearity, specificity, sensitivity, reproducibility, ruggedness, etc. Unfortunately, there is a tendency on the part of quality assurance departments to resist any method that cannot be validated *exactly* like a HPLC method.

Various workers in the field of NIR have attempted to address this (unavoidable) part of generating an equation for a NIR analysis. A comprehensive review of published papers and conference proceedings from the recent past displays a number of NIR methods validated under ICH guidelines or application of these guidelines in general.

A number of papers concentrated on the analysis itself, with an eye to the ICH validation of that method before use in the field. Blanco et al. [35] discussed the determination of miokamycin in different dosage forms. The word *validation* is even in the title of the paper. As a European group, they would be concerned with ICH more than FDA regulations, but the similarities are greater than any differences. The same group also reported on a validated diffuse reflection NIR method [36]. In both papers, the word *validation* figures prominently.

Ritchie and Ciurczak [38] discussed a near-infrared transmission method for sustained-release, prescription analgesic tablets. The entire method development process was performed with validation in mind. Additional general aspects of validation were discussed by Lange et al. [39] in yet another paper. The specific products were deemphasized in favor of the actual validation process.

Qualitative analyses were addressed by Ritchie [40]. In this talk, the concept of "disposable" equations was proposed. The idea was to treat a clinical trial as a closed population. That is, the exact lot of the active dosage form and the exact lot of the placebo dosage form are considered, for the particular study, all existing tablets or capsules of those materials. A simple discriminant equation, often using principal component analysis, can be validated, used to determine correct labeling of the patient blisters, and then discarded. Since that exact data (time since manufacture, storage conditions, etc.) will truly never exist, the equation will only be used once. The implications for validation were that the validation report dropped from a 50- to 75-page document to a 4-page memo. This was formally presented by Ritchie et al. [41].

In various publications, the analytical method was given equal time with the validation scheme. Forbes et al. [42] presented a trace analysis method for residual isopropanol in loracarbef using NIR. The validation implications for trace analyses were examined in the paper. Trafford et al. [43] used reflectance NIR to determine the active ingredient in paracetamol tablets. Again, validation issues were discussed.

Roller [44] described the development and validation of a NIR transmission method for the analysis of capsules. A similar paper, for the analysis of tablets, was presented by Ritchie et al. at that same meeting [45].

A number of papers and presentations concentrated on the validation of either the method or the software. Workman and Brown [46, 47] presented a two-part series on the practice of multivariate quantitative analyses. These papers concentrated on the algorithms involved in validated NIR methods.

The number of papers addressing method validation specifically has also been growing. Ciurczak addressed the overall problem for all types of spectroscopic methods in the industry [48]. He points out that ICH and FDA guidelines favor separation methods and do not address qualitative and quantitative methods based on spectroscopy. Conzen [49] addressed NIR quantitative validations in the industry. He also tried to reconcile current regulations with spectroscopic validations.

Klapper discusses NIR as a validatable tool in pharmaceutical quality control [50], while Moffat et al. [51] presented an excellent paper on how best to meet the ICH guidelines on validation. They used a quantitative method for paracetamol in tablets as an example for each guideline.

A similar paper was published by Pudelko-Korner [52] using the reflection analysis of sennosides in the parent plant as a way to address process and quality control concerns for validation.

All this activity implies that NIR is emerging as an accepted legitimate analytical tool. No longer are analysts saying, "Let's see if NIR can do the job"; rather, they are implementing the technique at the loading dock, the bench, at-line, and in process. Simultaneously, they are asking, "How can I best achieve regulatory approval of this method?" The focus now is on determining the best approach to satisfy the agencies' criteria for release testing.

References

1. J. Taylor, *Quality Assurance of Chemical Measurements*, Lewis Publishers, Chelsea, MI, 1987.
2. C. M. Riley and T. Rosanski, eds., *Development and Validation of Analytical Method Programs in Pharmaceutical and Biomedical Analysis*, vol. 3, Pergamon Press, Elsevier Science, New York, 1996.
3. M. J. Green, A Practical Guide to Analytical Method Validation, *Anal. Chem. News Features*, 68, 305 (1996).
4. M. E. Swartz and I. S. Krull, *Analytical Method Development and Validation*, Marcel Dekker, New York, 1997.
5. J. Miller and J. Crowther, eds., *Analytical Chemistry in a GMP Environment*, John Wiley & Sons, New York, 2000.
6. International Validation Forum, *Validation Compliance Annual, 1995*, Marcel Dekker, New York, 1995.
7. C. C. Chan, Y. C. Lee, H. Lam, and X. M. Zhang, eds., *Analytical Method Validation and Instrument Performance Verification*, John Wiley & Sons, New York, 2004.
8. J. Ermer and J. H. McB. Miller, eds., *Method Validation in Pharmaceutical Analysis*, John Wiley & Sons, New York, 2005.
9. J. M. Chalmers and P. R. Griffiths, eds., *Handbook of Vibrational Spectroscopy*, John Wiley & Sons, Chichester, UK, 2002.
10. United States Pharmacopeia, 2000, Chapter 1225.
11. United States Pharmacopeia, 2012, Chapter 1119.
12. ASTM E1655-97, *Standard Practices for Infrared, Multivariate, Quantitative Analysis*.
13. ASTM E1866-97, *Standard Guide for Establishing Spectrophotometer Performance Tests*.
14. ASTM E1790-04, *Standard Practice for Near Infrared Qualitative Analysis*.
15. International Union of Pure and Applied Chemistry, *Guidelines for Calibration in Analytical Chemistry. Part 1. Fundamentals and Single Component Calibration*, 1998.
16. International Union of Pure and Applied Chemistry, *Guidelines for Calibration in Analytical Chemistry. Part 2. Multispecies Calibration*, 2004.
17. International Union of Pure and Applied Chemistry, *Guidelines for Calibration in Analytical Chemistry. Uncertainty Estimation and Figures of Merit for Multivariate Calibration*, 2006.

18. ICH Guidelines for Method Validation, *Q2(R1) Validation of Analytical Procedures: Text and Methodology*, 1996.
19. Center for Drug Evaluation and Research, *Guidance for Industry Process Validation: General Principles and Practices*, 2008.
20. Center for Drug Evaluation and Research, *Guidance for Industry PAT—A Framework for Innovative Pharmaceutical Development, Manufacturing, and Quality Assurance*, 2004.
21. European Medicines Agency, *Note for Guidance on the Use of Near Infrared Spectroscopy by the Pharmaceutical Industry and the Data Requirements for New Submissions and Variations*, 2003.
22. European Medicines Agency, *Guideline on the Use of Near Infrared Spectroscopy by the Pharmaceutical Industry and the Data Requirements for New Submissions and Variations*, 2009.
23. European Medicines Agency, Guideline on the Use of Near Infrared Spectroscopy by the Pharmaceutical Industry and the Data Requirements for New Submissions and Variations, 2012.
24. European Medicines Agency, Guideline on the Use of Near Infrared Spectroscopy by the Pharmaceutical Industry and the Data Requirements for New Submissions and Variations, 2014.
25. Australian Pesticides and Veterinary Medicines Authority, *Guidelines for the Validation of Analytical Methods for Active Constituent, Agricultural and Veterinary Chemical Products*, 2004.
26. Good Automated Manufacturing Practice (GAMP) Forum, March 1998.
27. R. McGonigle, in *Development and Validation of Analytical Method Programs in Pharmaceutical and Biomedical Analysis*, vol. 3, ed. C. M. Riley and T. Rosanski, Pergamon Press, Elsevier Science, New York, 1996.
28. G. E. Ritchie and E. W. Ciurczak, Validating a Near-Infrared Spectroscopic Method, presented at Proceedings of Spectroscopy in Process and Quality Control Conference, New Brunswick, NJ, October 1999.
29. N. Draper and H. Smith, *Applied Regression Analysis*, John Wiley & Sons, New York, 1998, p. 69.
30. H. Mark, G. E. Ritchie, R. W. Roller, E. W. Ciurczak, C. Tso, and S. A. MacDonald, Validation of a Near-Infrared Transmission Spectroscopic Procedure. Part A. Validation Protocols, *J. Pharm. Biomed. Anal.*, 28, 251 (2002).
31. G. E. Ritchie, R. W. Roller, E. W. Ciurczak, H. Mark, C. Tso, and S. A. MacDonald, Validation of a Near-Infrared Transmission Spectroscopic Procedure. Part B. Application to Alternate Content Uniformity and Release Assay Methods for Pharmaceutical Solid Dosage Forms, *J. Pharm. Biomed. Anal.*, 29, 159 (2002).
32. M. Blanco, M. Bautista, and M. Alcalà, API Determination by NIR Spectroscopy across Pharmaceutical Production Process, *AAPS PharmSciTech*, 9, 1130 (2008).
33. A. Peinado, J. Hammond, and A. Scott, Development, Validation and Transfer of a Near Infrared Method to Determine In-Line the End Point of a Fluidised Drying Process for Commercial Production Batches of an Approved Oral Solid Dose Pharmaceutical Product, *J. Pharm. Biomed. Anal.*, 54, 13 (2011).
34. R. G. Whitfield, Near-Infrared Reflectance Analysis of Pharmaceutical Products, *Pharm. Manuf.*, 31 (1986).
35. M. Blanco, J. Coello, A. Eustaquio, H. Itturriaga, and S. Maspoch, Development and Validation of Methods for the Determination of Miokamycin in Various Pharmaceutical Preparations by Use of Near Infrared Reflectance Spectroscopy, *Analyst*, 124(7), 1089 (1999).

36. M. Blanco, J. Coello, A. Eustaquio, H. Iturriaga, and S. Maspoch, Development and Validation of a Method for the Analysis of a Pharmaceutical Preparation by Near-Infrared Diffuse Reflectance Spectroscopy, *J. Pharm. Sci.*, 88(5), 551 (1999).

37. D. McCune and N. Broad, Potency Assay of Antibiotic Tablets by Reflectance and Transmission NIR, presented at Proceedings of Eastern Analytical Symposium, Somerset, NJ, November 1999.

38. G. E. Ritchie and E. W. Ciurczak, Near-Infrared Transmission (NIT) Spectroscopy Assay of Continuous Release Prescription Analgesic Tablets, presented at Proceedings of Eastern Analytical Symposium, Somerset, NJ, November 1999.

39. A. J. Lange, Z. Lin, and S. M. Arrivo, Pharmaceutical Applications of Near-IR Spectroscopy, presented at Proceedings of Eastern Analytical Symposium, Somerset, NJ, November 1999.

40. G. E. Ritchie, Qualitative Near-Infrared Analysis in the Pharmaceutical Sector, presented at 9th International Diffuse Reflectance Conference, Chambersburg, PA, August 1998.

41. G. E. Ritchie, L. Gehrlein, and E. W. Ciurczak, Simultaneous Development, Validation and Implementation of a Near-Infrared (NIR) Diffuse Reflectance Spectroscopic Identification Method for Pharmaceutically Active and Inactive (Placebo) Clinical Dosage Forms, presented at Proceedings of PittCon, New Orleans, March 2000.

42. R. A. Forbes, B. M. McGarvey, and D. R. Smith, Measurement of Residual Isopropyl Alcohol in Loracarbef by Near-Infrared Reflectance Spectroscopy, *Anal. Chem.*, 71(6), 1232 (1999).

43. A. D. Trafford, R. D. Jee, A. C. Moffat, and P. Graham, A Rapid Quantitative Assay of Intact Paracetamol Tablets by Reflectance Near-Infrared Spectroscopy, *Analyst*, 124(2), 163 (1999).

44. R. W. Roller, Intact Capsule Content Uniformity Feasability by NIR Solid Dose, presented at Proceedings of PittCon, New Orleans, March 2000.

45. G. E. Ritchie, E. W. Ciurczak, and H. L. Mark, Validation of a Near-Infrared Transmission (NIT) Spectroscopic Assay of Sustained Release Prescription Analgesic Tablets, presented at Proceedings of PittCon, New Orleans, March 2000.

46. J. Workman and J. Brown, A New Standard Practice for Multivariate Quantitative Infrared Analysis—Part I, *Spectroscopy*, 11(2), 48 (1996).

47. J. Workman and J. Brown, A New Standard Practice for Multivariate Quantitative Infrared Analysis—Part II, *Spectroscopy*, 11(9), 24 (1996).

48. E. W. Ciurczak, Validation of Spectroscopic Methods in Pharmaceutical Analyses, *Pharm. Tech.*, 22(3), 92 (1998).

49. J. P. Conzen, Method Validation in Quantitative Near-IR Spectroscopy, *GIT Labor-Fachz.*, 42(2), 97 (1998).

50. A. Klapper, Use of NIR-Spectroscopy in Pharmaceutical Quality Control in Harmony with Official Specifications, *Pharma Technol. J.*, 19(1), 81 (1998).

51. A. C. Moffat, A. D. Trafford, R. D. Jee, and P. Graham, Meeting the International Conference on Harmonisation's Guidelines on Validation of Analytical Procedures: Quantification as Exemplified by a Near-Infrared Reflectance Assay of Paracetamol in Intact Tablets, *Analyst*, 125(7), 1341 (2000).

52. C. Pudelko-Korner, Quantitative Near-Infrared-Reflection-Spectroscopy of Sennosides in *Sennae fructus angustifoliae* in Processing and Quality Control as well as Validation of These Methods, *Pharma Technol. J.*, 19(1), 57 (1998).

chapter 7

Medical applications

The physics of near-infrared (NIR) spectroscopy are favorable for biological applications, especially for in situ measurements. The low absorptivities inherent in NIR allow the radiation to penetrate deeper, resulting in longer pathlengths. The sources are intense, giving more radiation to work with; detectors are sensitive and nearly noise-free, giving a more precise and accurate spectrum. The detector and lamp provide a sensitivity advantage even at very low light fluxes. These physical realities allow for measurements through tissue, muscle, fat, and body fluids with great precision. Fiber optic probes complete the picture and make NIR spectroscopy a technique adaptable to any lab or clinic.

The various topics in this chapter are listed by clinical application: blood chemistry, tissues, cancer, and so forth. Certain topics may be difficult to assign to a particular clinical application. In such cases, the authors' decision was to place the referenced paper in the most general application available. References to blood glucose, for instance, might appear in the sections headed "Blood Glucose" (7.1), "Review articles (7.9)," and "Blood Chemistry" (7.5). With hundreds of papers on each subject, we have chosen a cross section of the topics. This chapter is not intended to represent every paper published on the subject.

7.1   Blood glucose

One of the most publicized and pursued uses of near infrared in the life sciences recently is for in situ glucose measurements. Approximately 25,800,000 patients are believed to suffer from diabetes in the United States alone, with approximately 1.9 million new cases diagnosed in 2010. With the diagnostic market estimated in the multi-billion-dollar range, the number of scientists seeking a spectroscopic solution to this analytical problem is significant. A large number of patents for blood glucose measuring devices relying on near-infrared spectroscopy have been issued. Examples of the patented technology are as follows:

- The patented device developed by Ham and Cohen [1]. In this transmission device, the light is passed through the finger. A neural network (NN) algorithm recognizes the spectral areas of greatest correlation to glucose concentration and builds the calibration equation automatically. It is hypothesized that because of the complex nature of blood chemistry, complex algorithms will likely be needed for determination of any component of the blood.

• The patented device developed by Acosta et al. [2]. In this reflectance/transflectance device, the glucose content is correlated to redistribution of fluids between vascular and extravascular and intra- and extracellar compartments.

• The patented device developed by Acosta et al. [3]. In this reflectance device, the glucose content is directly modeled from the collected spectra. Reference values come from glucose testers available to diabetic patients.

• The patented methodology developed by Chung et al. [4]. The proposed method uses a ratio of absorbance measurements at select wavelengths to develop a model that correlates directly the glucose content to the spectral information.

These few examples of patented technologies and methodologies point to the various approaches that have been selected by researchers to predict blood glucose content—direct approaches that try to directly relate the glucose content to the NIR spectral information and indirect approaches that use a biological response to various glucose levels and attempt to relate these responses (more readily measurable by NIR) with glucose content. Along with the two types of approaches, significant work on chemometrics has been performed to enhance models' performance and robustness.

Modeling of the blood glucose system has been an ongoing project for numerous researchers for several decades. In 1999, Khalil [5] stated in a review document that to date, no experiment proved that the signal measured by near-infrared spectroscopy was the actual blood glucose concentration. Gary Small and Mark Arnold (University of Iowa), in particular, have published widely on the subject. In a 2005 review article [6], they surveyed the literature and discussed the merit of direct and indirect approaches to glucose modeling. Indirect glucose measurement is commonly based on the effect of glucose on the scattering properties of the tissue [7, 8]. Arnold and Small note that in indirect measurements, the major challenge is to ensure the selectivity of the measurement. Selectivity, along with sensitivity, is also an issue with direct measurements. Glucose is present at millimolar concentration levels, and while it does possess a unique vibrational spectrum compared to the other major blood components, existing devices may not have sufficient sensitivity to the analyte to develop robust tools.

Nevertheless, many authors have published on methods to predict blood glucose. In 1993 [9], Arnold and Marquardt published a paper modeling the NIR measurement of glucose in a protein-containing matrix.

The region from 2000 to 2500 nm was used for analysis of a series of glucose solutions ranging from 1.2 to 20.0 mM in a phosphate buffer closely resembling blood. The buffers contained such materials as bovine serum albumin (BSA). They found that a Gaussian-shaped Fourier filter combined with a partial least-squares (PLS) regression delivered a reasonable standard error of calibration. In all cases, the glucose absorbance at 2270 nm gave the optimal correlation, with a standard error of 0.24 mM. In 1994, the same authors reported work on a temperature-insensitive glucose measuring method [10]. In an approach similar to that just detailed [9], using a Fourier filter–PLS combination, a temperature range between 32 and 41°C was investigated. The temperature fluctuations caused relatively large variations in the spectra due to the water band shifts. Fourier filtering effectively eliminated the differences, providing a standard error that was even better than for their previous work (0.14 mM versus 0.24 mM). The necessity of spectral pretreatment is clearly seen in this study. Such temperature correction is necessary should measurements be taken from someone with a fever or who is in shock. This particular point is emphasized in many of the patents for analytical devices. A system to ensure consistency of the temperature of the skin at the sampling site is often included so that interferences due to temperature are limited.

Arnold, Small, and colleagues performed further work in 1996, this time using physiological levels of glucose in the presence of protein and triglycerides [11]. The solutions used in the study contained between 1 and 20 mM glucose. The interferences were varied within each level of glucose. It was seen that multivariate algorithms compensated for the chemical variations in the blood where the glucose level remains unchanged. The same researchers also performed some published work on the use of quadratic PLS and digital filtering techniques to account for non-glucose-related changes in the spectra [12]. They concluded that pretreatment helped reduce interferences and resulted in more robust equations. The instrumentation (either single-beam or double-beam) is made less important as interferences are accounted for in the NIR equation.

Two interesting papers by the research team of Small and Arnold were published in 1998 [13, 14]. They were devoted to the calculations used in noninvasive blood glucose measurements. The authors investigated neural networks and partial least-squares as calibration approaches and examined such things as signal-to-noise enhancements through an understanding of how light is attenuated as it passes through tissue. The companion articles go into detail regarding the scattering effects of blood and tissue. Compensation schemas are identified and proposed to alleviate some of these interferences. It is strongly recommended that fat, water, and tissue be compensated for in any model considered. Over a dozen salient references are listed.

David Haaland et al. added to the modeling literature with a 1992 paper [15]. This work used whole blood for the model. Scanning from 1500 to 2400 nm, a PLS equation was developed on glucose-spiked whole blood. The range between 0.17 and 41.3 mM yielded an equation with a standard error of 1.8 mM. Four patients were used as models for this project. Cross-validated PLS standard errors for glucose concentration based on data obtained from all four subjects were 2.2 mM. When PLS models were developed on three patients' blood samples and tested on the fourth, the glucose predictions were poor. The conclusion was that models must be developed for individual patients due to intersubject variability of blood chemistry.

Another novel device has been developed in Germany by Schrader, where a laser illuminates the humor of the eye and the resulting absorbance spectrum is used to measure the amount of blood glucose in the patient [16]. The device is based on a patent developed by Backhaus et al. [17]. He found that the glucose levels in the anterior chamber of the eye closely follow changes in blood glucose with a latency of approximately 20 min. The equations developed by this instrumentation allow for noninvasive monitoring of physiological glucose levels with an error of ±30 mg/dl. The basis for using NIR for determination of blood glucose by scanning through skin and muscle is that the blood glucose level in the blood is similar, if not identical, to the glucose level in tissues. This was claimed by Fischer et al. in 1994 and later demonstrated through a series of measurements [18].

However, contradicting this work was a paper by Sternberg et al. [19]. This group claimed that tissue contained only 75% of the glucose level found in blood. Fortunately, NIR measurements are inclusive of blood and tissue. The calibration was based on a point of contact for each individual patient; thus, the ratio of tissue to venous/arterial blood will be a constant. Correlation of the spectra to blood glucose readings is then acceptable.

In order to model more correctly the in vivo realities of human body chemistry, "phantoms" (simulated biological conditions, containing fat, blood, skin, etc.) were built for simulated in vivo testing. Arnold et al. built phantoms of water, fat, and muscle tissue to mimic the skin of a patient [20]. They found that in vivo overtone spectra collected across human webbing tissue with a thickness of 6.7 mm could be simulated with a water layer thickness from 5.0 to 6.4 mm combined with a fat layer thickness from 1.4 to 4.2 mm. For the purposes of this study, animal tissue and fat were used; there is little difference in composition between human and animal materials. They concluded that these phantom studies would help researchers develop patient-applicable methods.

This phantom work was continued by Arnold et al. in a later publication [21]. This is a negatively designed study, used as an object lesson to warn the inexperienced user about the pitfalls of chemometrics. They used an in vitro model of blood-simulated samples to build a model for blood glucose determination. In this case, however, they carefully omitted any glucose from the samples. The samples were randomly assigned glucose values and a PLS regression developed. As with any PLS model, an equation was created to provide reasonable standard errors, regression coefficients, etc. Since there was no glucose present, this equation could not predict glucose when samples containing the sugar were tested. This paper displayed potential errors to avoid during calibration development and analysis.

An interesting observation made by Maier et al. was that there exists a correlation between blood glucose concentration and the reduced NIR scattering coefficient of tissue [22]. Using a frequency domain NIR spectrometer, the scattering coefficient of tissue was measured with adequate precision to detect changes in glucose. The work was based on the theory that as the glucose concentration increased, the refractive index of the blood also increased in a predictable manner. This increased refractive index would then decrease the scattering coefficient of the blood and give an indication of the concentration. Some questions of the applicability of this work to in vivo measurements must be asked, but it does demonstrate one of the novel approaches being investigated in the field.

Some researchers working at the University of Krakow have published several papers and given numerous talks and posters devoted to the mathematical treatments of the complex spectra produced from the NIR examination of blood through skin and muscle [23–25]. They have been working with neural networks (NNs) in particular and have made some interesting observations. The advancement and proliferation of work in this field may be traced directly to more powerful personal computers. These complex algorithms simply could not be run on desktop computers a few decades ago.

One other technique [26] uses a fiber optic light pipe to measure blood glucose through the skin of a finger. The device uses a portion of the fiber, stripped of its cladding, as a virtual attenuated total reflectance (ATR) device positioned against the skin of the thumb. Because so much radiation is lost into the skin, white light was used with a monochromator situated postsample. This configuration gave better sensitivity than when first resolving the wavelengths. Postsampling indicates that "white" light is impinged on the skin, and that the resultant emerging light is collected and submitted to a monochromator. A study by the team of Dr. Ozaki [27] investigated the use of a particular illumination and detection geometry with fiber optics that allowed them to successfully interrogate dermis tissue and reduce other interferences. Using partial least-squares, they were able to obtain prediction error for all the subjects involved in the study of 23.7 mg.dl, thus showing similar results to other studies and the possibility to use the same model for different patients.

One nonskin application was published in a paper by Heise et al. [28], where a procedure for measuring blood glucose through the lip was described. They used the 1100 to 1800 nm wavelength range, and partial least-squares was the algorithm of choice. The mean square prediction error was estimated between 45 and 55 mg/dl. A lag time of approximately 10 min existed between the drawn blood values and the values derived from the lip tissue. The authors recommended using fiber optics for further developments in this field.

An example of indirect measurement of blood glucose was provided by the team of Small and Arnold in a study investigating the change in blood scattering with respect to the change in concentration of glucose [29]. Authors showed prediction errors in the 1 to 2.5 mM range, depending on the part of the NIR signal considered.

A lot of work has been put into the development of chemometrics methods that would extract the most relevant signal to allow an improved prediction performance and robustness of the glucose models. Besides the work done with NN, literature exists on the use of the net analyte signal (NAS). Initially developed by Lorber as a means to calculate multivariate figures of merit [30], the NAS corresponds to the net signal of the analyte of interest. It can be used as a preprocessing step to focus the analysis on only the relevant signal. An example of the use of NAS for blood glucose can be found in Ren and Arnold [31]. On the same topic of removing interfering signal, techniques based on eliminating the signal that is not related to the signal of glucose have been published. Thus, the team of Ozaki et al. proposed to use a local orthogonal signal correction method to remove interferences from the main blood components in addition to searching for the best suited wavelength regions with a moving window partial least-squares [32, 33]. Another technique utilizing the uninformative variable elimination combined with successive projections was intended by Li et al. [34]. A nonlinear regression version of the uninformative variable elimination was later proposed by the same authors [35].

To improve model robustness and performance, several authors have worked at optimizing the samples included in calibration. So et al. [36] used a Monte Carlo approach to determine the best set of samples that should be used. The use of simulated spectra has also been tested with results in in vivo experiments in the 12.3 mg/dl range [37].

7.2   Blood oxygenation

Venial and cranial blood oxygenation is a simultaneously (relatively) simple and yet nontrivial measurement. Early reports of NIR for diagnostic applications came from researchers such as Jobsis in 1977 [38]. He used NIR to monitor the degree of oxygenation of certain metabolites. Later, Ozaki et al. [39] examined venal blood to determine the level of deoxyhemoglobin. The back of the hand was illuminated, and the diffusely reflected light was captured by a miniature integrating sphere equipped with a PbS (lead sulfide) detector. The spectra were correlated with results from a CIBA Corning 278 blood gas analyzer. The 760 nm band in the spectrum was seen to correlate quite well with deoxyhemoglobin, and a negative correlation exists with oxygenated hemoglobin.

Michael Sowa and his group [40] used NIR imaging as a noninvasive technique to monitor regional and temporal variations in tissue oxygenation. The purpose was to ascertain the effects of restricted blood outflow (venous outflow restriction) and interrupted blood inflow (ischemia). In this work, the software was the heart of the paper. Multivariate analyses of image and spectral data time courses were used to identify correlated spectral and regional domains. Fuzzy C-means clustering of image time courses revealed finer regional heterogeneity in the response of stressed tissues. The wavelength region from 400 to 1100 nm was monitored from 0 to 30 min, and a plot of these data was developed to produce a topographical representation of the phenomenon. Peaks and valleys were apparent where blood became oxygenated and deoxygenated. These standard wavelength-based values correlated well with the images developed by the 512 × 512 back-illuminated charge-coupled device (CCD) element.

Clustering results clearly showed areas of both low and high oxygenation. These results have important implications in the assessment of transplanted tissue viability. Mancini et al. [41] estimated skeletal muscle oxygenation by using the differential absorption properties of hemoglobin. Oxygenated and deoxygenated hemoglobin have identical absorptivities at 800 nm, while deoxygenated hemoglobin predominates at 760 nm. Effects of myoglobin on the readings were also investigated, and it was found that the readings' correlations were due to hemoglobin oxygenation. Venous oxygen saturation and absorption between 760 and 800 nm were correlated. Mancini and colleagues reached several conclusions: (1) hypoxia in KCl-arrested hearts results in only moderate activation of anaerobic glycolysis, (2) oxygenation of the epicardial and midmural layers is similar, and (3) a large pO_2 gradient exists between vascular and intercellular space in beating and arrested crystalloid-perfused hearts.

Lin and York examined the influence of fat layers on the determination of blood oxygenation in 1998 [42]. The phantom experiments showed fat influences patient-to-patient measurements. This is more easily compensated for in any individual patient. Yamamoto et al. addressed the issue of fat interference with an oximeter that corrected for the influence of subcutaneous fat [43]. The wavelengths, again, were the key, as was the algorithm.

The effect of water on NIR determination of hemoglobin concentration in a tissue-like phantom was studied by Franceschini et al. in 1996 [44]. Their in vitro studies consisted of aqueous suspensions containing

Liposyn, bovine blood, and yeast, buffered at pH 7.2. The optical coefficients of the mixture matched those of biological tissue in the NIR, and the hemoglobin concentration (23 μM) was also similar to that found in tissues. They oxygenated and deoxygenated the hemoglobin by sparging the mixture with either oxygen or nitrogen. They determined that water concentration must be taken into account to obtain accurate results of hemoglobin concentrations.

Z. X. Jiang et al. presented a device that allowed noninvasive measurement of cerebral tissue oxygenation to be performed [45]. Again, based on fiber optics, shorter NIR wavelengths were used. Another device allowing diffuse reflectance measurements of the skin was developed by Marbach and Heise [46]. The device presented has an on-axis ellipsoidal collecting mirror with efficient illumination for small sampling areas of bulky body specimens. The actual schematic is too complex to describe in this chapter. The researchers supported the optical design with a Monte Carlo simulation study of the reflective characteristics of skin tissue. While their work was centered on the 1600 nm peak associated with glucose (using the lip as the point of entry), the work is applicable to other tissue research.

Keiko Miyasaka presented some of his work [47] at a meeting in Toronto. As a worker in the field of critical care for children, he introduced what he calls a Niroscope for near-infrared spectroscopy. His work was performed during pediatric anesthesia and intensive care. Dr. Miyasaka found the Beer's law relationship not followed rigorously when the signal was passed through the cranium. Considering the massive scattering absorbed light, this was understandable. It did, however, give a semiquantitative or indicating equation. What Miyasaka was measuring was the intercranial levels of oxygenated hemoglobin (HbO_2), deoxygenated or reduced hemoglobin (Hb), and cytochrome redox status. Two methods were used: photon counting and a micro-type pulse laser. The photon counting method is necessary because of the extreme attenuation of the incident radiation when transversing the cranium. The pulse laser was used to enhance the amount of light introduced into the brain.

Three conclusions may be reached from the NIR data: (1) changes in HbO_2 levels reflect changes in arterial blood, (2) Hb changes are due to venous blood, and (3) total hemoglobin reflects changes in cerebral blood volume or intercranial pressure. This tool will be invaluable for emergency and operating room situations for both children and, someday, adults. Van Huffel and Casaer used NIR to monitor brain oxygenation. They used the information to correlate with behavioral states of preterm infants and to understand the development of brain hemodynamics autoregulation [48]. The concentrations of HbO_2, Hb, and cytochrome aa3 (Cytaa3) are used to monitor the oxygenation level in infant brain blood. Some novel chemometrics were involved as well; windowed fast Fourier transform (WFFT) and wavelet analyses were employed. The purpose of the work was to see relationships between the computed chromophore concentrations and heart rate, breathing, and peripheral oxygen saturation. They presented similar work in 1998 as well [49].

Chris Cooper et al. [50] performed another study; this one aimed at the adult brain. In this work, NIR was used to determine the effects of changes in the rate of oxygen delivery on adult rat brain chemistry. Absolute levels of oxyhemoglobin, deoxyhemoglobin, and the redox state of the CuA center in mitochondrial cytochrome oxidase were determined. An interesting finding was that as the mean arterial blood pressure reached 100 mmHg, hemoglobin oxygenation began to fall, but the oxidized CuA levels only fell when cerebral blood volume autoregulation mechanisms failed at 50 mmHg. Hemoglobin oxygenation fell linearly with decreases in the rate of oxygen delivery to the brain, but the oxidized CuA concentration did not start to fall until this rate was 50% of normal. The results suggested that the brain maintained more than adequate oxygen delivery to mitochondria. Their conclusion was that NIR is a good measure of oxygen insufficiency in vivo.

A related study on human infants was performed by J. S. Wyatt et al. [51]. They used NIR to quantify the cerebral blood volume in human infants using NIR spectroscopy. Similar difficulties were encountered with the amount of light actually penetrating the cranial cavity, but useful equations were generated.

Kupriyanov et al. determined intracellular pO_2 in cardiac muscle by the balance between its diffusion from vascular to intercellular space and its uptake by mitochondria [52]. They reasoned that cessation of mechanical work decreased O_2 demand and should have reduced the O_2 gradient between vascular and intercellular spaces. For their research, they compared the effects of arterial pO_2 on myoglobin (Mb) oxygenation, O_2 uptake, and lactate formation rates in beating and KCl-arrested pig hearts.

Ischemia in the forearm was studied by Mansfield et al. in 1997 [53]. In this study, the workers used fuzzy C-means clustering and principal component analysis (PCA) of time series from the NIR imaging of volunteers' forearms. They attempted predictions of blood depletion and increase without a priori values for calibration. For those with a mathematical bent, this paper does a very nice job describing the theory behind the PCA and fuzzy C-means algorithms.

Another interesting paper was published by Wolf et al. in 1996 [54] where they used NIR and laser Doppler flowmetry (LDF) to study the effect of systemic nitric oxide synthase (NOS) inhibition on brain oxygenation. The study, performed on rats, demonstrated no effects on brain oxygenation during cortical spreading depression (CSD).

Doppler ultrasound was combined with NIR in another study [55]. Liem et al. used NIR and ultrasound to follow the cerebral oxygenation and hemodynamics in preterm infants treated with repeated doses of indomethacin. In addition to the normal concentrations of oxyhemoglobin, deoxyhemoglobin, and oxidized cytochrome aa3 measured by NIR, transcutaneous pO_2 and pCO_2, arterial O_2 saturation, and blood pressure were measured as well. Along with the cerebral blood volume, they were all used for diagnosis and research. Low oxygenation was then thought to be a possible contraindication for indomethacin treatment for preterm infants.

The physical placement of detectors on the scalp for brain blood oxygenation was studied by Germon et al. in a 1998 study [56]. Detectors placed 2.7 and 5.5 cm from a NIR emitter were compared for the determination of Hhb, O_2 Hb, oxidized cytochrome C oxidase, and total hemoglobin. The biological portion of the experiment was to measure the chemical changes with an induced reduction of the mean decrease in middle cerebral artery blood flow. The signal change per unit photon pathlength detected at 5.5 cm, for Hhb, was significantly greater than at 2.7 cm. On the other hand, the increases in all chromophores detected at 5.5 cm during scalp hyperemia were significantly less than those detected at 2.7 cm. More work is indicated before meaningful applications can be designed from this work.

Using similar instrumentation, Henson et al. determined the accuracy of their cerebral oximeter under conditions of isocapnic hypoxia [57]. Using healthy volunteers, dynamic endtidal forcing was used to produce step changes in positron emission tomography (P_{ET}) O_2, resulting in arterial saturation ranging from ~70 to 100% under conditions of controlled normocapnia (resting PT_{ET} O_2) or hypercapnia (resting plus 7–10 mmHg.) Using standard methods, the O_2 concentrations for each patient under each condition were determined. Excellent correlation resulted in the rSO_2 and $[S_{circ}]_{j[vbar]}O_2$ for each individual patient. However, wide variability between patients was discovered. They concluded that under the current limitations of the equipment, the device was good for tracking trends in O_2 but could not be used as an absolute measure for different patients.

Numerous and disparate studies have been published or presented in recent years regarding the effects of various normal and pathologic conditions on blood oxygen:

1. Hoshi et al. investigated the neuronal activity, oxidative metabolism, and blood supply during mental tasks [58].
2. Okada et al. presented work on impaired interhemispheric integration in brain oxygenation and hemodynamics in schizophrenia [59].
3. Hoshi et al. looked into the features of hemodynamic and metabolic changes in the human brain during all-night sleep [60].
4. Akına et al. studied the clinical application of NIR in migraine patients [61]. They assessed the transient changes of brain tissue oxygenation during the aura and headache phases of a migraine attack.

Surgeons are concerned with brain blood flow to patients undergoing cardiopulmonary bypass surgery. An intensive study by Chow et al. was conducted where blood flows were restricted to patients from age 2 weeks to over 20 years [62]. Near infrared was used to correlate blood flow rate with NIR spectra of the brain. Flows of 0.6, 1.2, and 2.4 L/m^2/min were used. Their results showed that flow was related to mean arterial pressure, but did not correspond to pulsatility. This was interesting in that pulse rate is often used as a diagnostic to assure sufficient blood flow to the brain during surgery.

Totaro et al. published a detailed paper on the factors affecting measurement of cerebrovascular reactivity when measured by NIR [63]. Some of the points covered were the relative transparency of the skin, skull, and brain in the 700 to 1100 nm region and the oxygen-dependent tissue absorption changes of hemoglobin. Their study covered all relevant factors, such as age, sex, reproducibility (often neglected in many academic papers), and venous return. The test was based on a 3 min baseline, a 3 min hypercapnia (5% CO^2 in air), and a 2 min recovery period.

Changes in NIR spectra and transcranial Doppler sonography parameters were significantly correlated with variations of end-tidal CO_2 ($p < 0.005$). In addition, a significant correlation between the reactivity indexes (approximately, absorptivities) of NIR spectrometry parameters and flow velocity was found ($p < 0.01$). High reproducibility was also found for deoxyhemoglobin ($r = 0.76$), oxyhemoglobin ($r = 0.68$), and flow velocity ($r = 0.60$) reactivity indexes. No significant differences between the reactivity indexes of different body positions were found ($p < 0.05$). The reactivity indexes of oxyhemoglobin decreased ($p > 0.05$) and deoxyhemoglobin increased ($p < 0.01$) with age. Their overall conclusion was that NIR is a viable technique for evaluation of cerebrovascular reactivity for patients with cerebrovascular disease.

Some exciting work was reported by Hitachi at a recent meeting in Japan [64]. The research, conducted at the Tokyo Metropolitan Police Hospital, used NIR to detect blood flow changes in the brain to determine sites of epileptic activity. The location of blood flow increases corresponded well with conventional methods such as intercranial electroencephalogram (EEG) or single-photon-emission computed topography (SPECT). The technique was able to determine the side of the brain where the episode was taking place in all the patients on which it was tried. This technique could replace the intrusive electrodes currently in use. Hitachi plans to expand this technique to other brain diseases.

Near-infrared spectroscopy has been used to determine the activity of the brain using oxygenation levels. More particularly, a series of articles by Boas et al. [65, 66] discusses the limitations that single-point NIR has in the analysis of focal brain activation compared to diffuse activation. Researchers found that it is possible using NIR spectroscopy to determine regions of the brain responsible for simple motor tasks such as moving the finger. However, the local nature of the focal activation is a challenge to NIR measurements. The main source of error was determined to be the cross talk between the pathlength of the entire area illuminated and the partial pathlength of the activated region, thus making it difficult to perform accurate measurements of the focal change. Authors determined that some wavelengths were more prone to cross talk (780 and 830 nm pairing compared to 690 or 760 nm pairing with 830 nm).

On the same topic, a significant amount of work has been done to study the level of oxygenation of the brain during exercise at submaximal [67] and maximal [68, 69] levels. In a review by Rooks et al. [70], some very interesting results provided by NIR showed that oxygenation of the brain increases during exercise to hit a plateau and then decline toward a baseline level at very hard exercise levels. The response was modified as a function of the training level, with lower oxygenation levels attained by less fit people. Interestingly, the review ends by stating that improvements of NIR should allow a better understanding of local focal changes, a task undertaken a decade earlier by Boas et al. [65].

Cerebral oxygenation has been used in a multitude of other applications and helped develop the field of functional near-infrared spectroscopy (fNIRS). In addition to the topics already mentioned, fNIRS has found application in speech analysis and disorder [71], psychiatry and disease state [72], cerebral ischemia and hypoxia [73, 74], brain injuries [75, 76], and cardiac surgery [77, 78], to only name a few. However, in a review, Highton et al. [79] noted that fNIRS is limited by the contamination of the signal by extracranial tissues. In addition, the large variability between individuals makes it difficult to set thresholds over or under which a problem is detected, thus making fNIRS ideal for trending and monitoring, but limiting decision making to the individual level.

Improvement in the specificity of the measurements is now the focus of the research. Examples of measurements of nerve oxygenation have been presented [80].

Finally, a significant amount of work has been done with optoacoustic systems to image the vascular system. Using short pulses of near-infrared light, strong absorbers such as hemoglobin can be used to provide detailed images of blood vessels and are well suited for diagnosing and monitoring tissue pathologies such as those induced by tumors [81–84]. By modulating the wavelengths used, it is then possible to model the blood oxygen saturation and provide cancer researchers with the possibility to monitor tumor oxygenation [85].

In summary, oxygenation in general has been a major focal point for NIR medical research [86–112]. It continues to be a successful application for NIR.

7.3 Tissue

Dreassi et al. published a series of papers on atopy of skin. The first [113] discusses how NIR penetrates complex structured matrices to at least 0.20 mm. He found NIR to give valuable insights into the stratum corneum. Using principal component analysis, the team decomposed the global structural information into components such as water and lipid structures. In another paper, the group studied interactions between skin and propylene glycol 400 (PEG 400), isopropyl myristate (IPM), and hydrogel [114]. They examined spectral phenomena correlated with varied water and lipid content in normal and atopic skin after reaction with these reagents. These particular chemicals were chosen to represent a prevalently hydrophilic solvent (PEG 400), a prevalently lipophilic solvent (IPM), and a hydrophilic pharmaceutical gel used to promote contact in electrocardiography. By using principal component analysis of the NIR spectra, they were able to distinguish atopic from normal skin after simple contact with these reagents. Similar results were reported in a later work from this group [90] using a series of perfluorinated polyethers (fomblins) of differing molecular weights and viscosities. Interaction between the chemicals and the organ (skin) involves two stages. First, the skin is physically modified. Second, the water is moved and redistributed. It was assumed that the chemical agents caused changes in the stratum corneum. The nature of the changes differed between normal and atopic skin. Different mechanisms appeared to operate in each case, and each of the chemicals led to spectral differences between atopic and normal skin.

One important assessment made with NIR is the viability of tissue after trauma [115, 116]. Prolonged and severe tissue hypoxia results in tissue necrosis in pedicled flaps. The group used NIR to identify tissue regions with poor oxygen supply. The work was performed on reversed McFarlane rat dorsal skin flaps. It was seen that oxygen delivery to the flap tissue dropped immediately upon the onset of atopy. As expected, severe trauma that causes severing of the skin from the main blood flow causes necrosis of the tissue. Near infrared may be used as a tool in assessing the success of reattachment of the traumatized skin. Viability of the skin as indicated by its oxygenation level has been investigated by Attas et al. [117]. Similarly to the oxygenation studies presented in the previous section, authors determined skin and blood parameters and used them as indication of health. Application of the measurements are tremendous in fields such as plastic surgery and clinical skin evaluation.

The viability of surgical flaps has also been investigated, and the detection of anastomosis thrombosis was found possible, allowing the salvage of flaps [118]. In the study, 48 patients corresponding to 50 flaps were continuously monitored by NIR. Authors were able to differentiate between arterial and venous thrombosis. In burn wounds, NIR has been used to quantify edema by measuring the water concentration changes [119, 120]. In control regions, the water content, measured after biopsy, had not changed, while in burned areas, it had increased by 18% on the superficial regions and by 23% for deep injuries. These measurements can also be useful in forensics where NIR has the potential to be used to determine the age of bruises based on the measurement of skin water and hemoglobin content [121].

Another tissue type—nails—was studied by Sowa et al., both in vivo and ex vivo [122]. Mid-IR (MIR) and NIR spectra were collected for viable and clipped human nails. Depth profiling by MIR was performed nonintrusively by photoacoustic spectroscopy (PAS). Near-infrared ATR, NIR diffuse reflectance, and PAS were compared. Band assignments were made, such as the N-H stretch–amide II bend combination centered at 4868 cm^{-1} in this basic study. They concluded that the lower-energy NIR-ATR, for purposes of their study, gave the best results.

Interesting measurements, such as body fat in infants, are easily made by NIR [123]. Newborn body development could be evaluated as to body fat addition due to breast-feeding versus non-breast-feeding nutritional input. This is a more accurate measurement of proper development than mere body weight, which includes other tissues and bone. Another application to pre- and newborns was published by Liu et al. in 1997 [124]. In this paper, they presented a measure of fetal lung maturity from the spectra of amniotic fluid. The lecithin/spingomyrlin (L/S) ratio was determined by thin-layer chromatography (TLC) and used to calibrate a NIR equation using the whole amniotic fluid extracted from pregnant women. About 350 µl of fluid was required. This was scanned from 400 to 2500 nm using a commercially available instrument. The correlation between further samples of fluid and TLC results was about 0.91, considered excellent for the complexity of the solution and extremely small sample size. A multivariate regression method was needed due to the complex nature of the samples.

The temperature of tissue was measured by Barlow et al. in 1995 [125]. Absorbance changes in the water spectrum between 700 and 1600 nm (in transmission) and the spectrum between 800 and 2200 nm (reflectance) were found to correlate with the temperature of the tissue in which the water is contained. The standard error of estimate (SEE = 0.02 to 0.12°C) and standard error of prediction (SEP = 0.04 to 0.12°C) were found. Since tissue in general is a highly dispersing medium, various attempts have been made to mitigate this scattering.

Tsai et al. presented a paper that was merely concerned about the absorption properties of soft tissue constituents [126]. They concentrated their work in the region from 900 to 1340 nm. In NIR, the shorter wavelength regions have lower absorptivities and, consequently, deeper penetration in tissue. At the same symposium, Schmitt et al. presented a paper wherein the processing of NIR spectra from turbid biological tissue was discussed [127]. Much energy has been used to obviate the scattering effects of tissue. Discrimination of the actual absorption of light versus losses due to scattering demands the use of higher-order algorithms.

Andersson-Engels et al. have presented a series of papers on time-resolved transillumination of tissue, specifically with tumor detection in mind [128–132]. In these papers, the group goes into detail about the physics involved in using a picosecond diode laser, a mode-locked argon ion-dye laser, or a mode-locked Ti-sapphire laser to conduct time-resolved spectroscopy on tissue. In one case, a human (female) breast is compressed to ~35 mm for the test. Light, in 100 fs pulses at 792 nm (giving a 50 ps apparatus function), is dispersed to a signal that is more than 1 ns long. The dispersion curve obtained contains information about the optical properties of the tissue. In the case of scattering-dominated attenuation (scattering coefficient >> absorption coefficient), detection of early trans-mittal light will be practically insensitive to variations in the absorption coefficient since it uses wavelengths nearly seen as baseline. The scattering properties determine the amount of detected early light. This is important for optical mammography for which neovascularization surrounding a tumor causes an increased light absorption in the tumor region. A model has been developed that accurately predicts the time dispersion curves obtained experimentally. This breakthrough should greatly aid spectral imaging in vivo.

Skin lesions have been the topic of several studies. McIntosh et al. [133] used NIR and discriminant analysis to group normal skin and skin lesions. Their method showed that it was possible to classify spectra between benign lesions and premalignant and malignant lesions. An in-depth spec-tral investigation of the difference between normal and cancerous skin was performed by Salomatina et al., who found that there was a significant dif-ference in the scattering properties of the skin to allow successful classifica-tion [134]. Also, short NIR wavelengths have been used in a hyperspectral imaging application for the detection of actinic keratosis [135].

The pioneering work in the assessment of arterial walls has been performed at the University of Kentucky Medical Center. Robert Lodder has been producing excellent results in this field [136, 137]. In these early papers, the location and quantities of high-density lipoprotein (HDL), low-density lipoprotein (LDL), and apolipoproteins in living tissue were determined. Their clinical purpose was to study intact arterial walls and compare normal and atherosclerotic patients. The technique appeared to allow monitoring of a number of molecules intimately associated with atherogenesis, including collagen and elastin, cholesterol and cholesterol esters, and calcium salts. The drawbacks to their work were the 30 min collection times and the 500 mW power needed for the studies. From this initial work, a significant amount of research has been performed on the topic with successful detection of plaque in arteries [138–143]. Catheter-based systems have been developed to image plaque without the need to flush the tissues [144, 145]. Plaque imaging has also been studied with a multitude of recent publications reporting encouraging results [146–148].

The detection of rheumatoid arthritis by NIR has also been tested. In a study by Canvin et al. [149], a relationship between spectral variance and joint tenderness (swelling) and radiographic damage was found, allowing the diagnosis of early and late stages of arthritis.

7.4   *Major organs*

Sato et al. studied hepatic oxygenation differences between total vascular exclusion and inflow occlusion [150]. The studies were performed under liver ischemia conditions. The oxygen content in the hepatic venous blood was measured by NIR. Hepatic oxygenation, microcirculation, and function were studied by Jiao et al. [151] in a recent paper. The blood flow in the liver and its function were also studied directly in a cirrhotic animal model by the same authors [152]. Peripheral blood clearance of indocyanine green was shown to be less accurate than direct NIR probes on the liver surface. Changes in tissue oxygenation of a porcine liver were measured using NIR by El-Dosoky et al. [153]. A laparotomy was performed and the surface of the liver exposed. Probes placed on the liver surface continuously measured changes in hepatic tissue oxyhemoglobin, deoxyhemoglobin, and the reduction-oxidation state of cytochrome oxidase.

Similar work was performed by Teller et al. [154]. Newborn infants' livers were measured during nasogastric tube feeding. The optical properties of liver tissue and liver metastases were examined by NIR in a paper by Germer et al. [155]. Since laser-induced thermotherapy is becoming popular, the team wanted to describe the light scattering effects of both types of tissue. More recently, NIR has been used to measure the liver oxygenation levels in ill children, thus reducing the need for invasive pulmonary artery catheters [156].

Renal failure was studied by Cochrane et al. [157]. They used ischemic preconditioning to attenuate functional, metabolic, and morphologic injury from ischemic acute renal failure in rats. The value of preconditioning was studied in particular. Numerous other studies describe the use of NIR for the measure of oxygenation of the kidney. Studies by Lane et al. and Vidal et al. looked at the possibility to monitor the oxygenation kinetics during kidney transplant [158, 159]. Other works were reviewed by Mayevsky et al. [160]. An interesting study investigated the possibility to use NIR for the evaluation of lower limb ischemia in patients undergoing dialysis [161]. The study found that after dialysis, the oxygenation level of the blood decreased compared to baseline levels, thus increasing the frequency of cramps and other physical discomfort.

An indirect measure of the pancreatic activity was performed by Faggionto et al. [162]. The team studied the fecal fat of 594 cystic fibrosis patients and determined that while a limited relation existed, it was possible to gain information about the degree of pancreatic inefficiency. Similar studies were performed by other research groups to gain information about pancreatic steatorrhea [163] and maldigestion and malabsorption [164, 165]. Sugar was added as an analyte to fat and nitrogen in the study performed by Rivero-Marcotegui et al. [166]. Yet another study of feces analyzed for fat, neutral sterols, bile acids, and short-chain fatty acids [167].

Gastric mucosal microcirculatory disorders in cirrhotic patients were studied by Masuko et al. [168]. The rheologic properties of the gastric microcirculation were determined with an endoscopic laser Doppler flowmeter. Bowel viability has also been investigated. During a jejunal autograph, Hirano et al. [169] used NIR to determine the tissue oxygen saturation of the graph. NIR was also found as a good alternative to gastric tonometry in infants [170].

Several groups have studied lung function using near infrared. Noriyuki et al. [171] studied the tissue oxygenation of lungs under several conditions: hypoxic loading, administration of an inhibitor (NaCN), and hemorrhagic shock. They found that changes in the hemoglobin oxygenation state in the lung depended on inspired oxygen concentration, NaCN-induced reduction of cytochrome oxidase aa3 was observed, and total hemoglobin levels decreased after bleeding. Tirdel et al. [172] researched metabolic myopathy as a cause of the exercise limitation in lung transplant recipients, while Noriyuki et al. [173] evaluated lung tissue oxygenation using NIR spectroscopy. Other researchers have used NIR as a tool for lung graft viability [174].

In summary, NIR is a tool of choice for the determination of good functioning of organs. As stated in a recent review by Tweddell et al. [175], "NIRS is 'standard of care' for postoperative management." A review by Kravari et al. [176] provides a good overview of the work done in medicine with NIR for the monitoring of the major organ's health. While not explicitly stated in this section, as it was investigated more in depth in the oxygenation section (7.2), NIR has been used to monitor cerebral oxygen saturation before heart surgery [177].

7.5 Blood chemistry

The term *blood chemistry* usually implies in vitro uses of NIR. As an example, cell culture media were analyzed by McShane and Cote in 1998 [178].

Samples of a 3-day fibroblast culture were analyzed by standard clinical techniques as well as by NIR. Glucose, lactate, and ammonia were determined after building a model from several lots of cell culture media. The purpose was to follow the nutrient levels to determine noninvasively when fermentation was complete. The approach was deemed successful.

Jeff Hall used NIR to analyze the major components of human breast milk [179]. This application could help nutritionists determine (quickly) whether a nursing mother needs supplements for her child. Additional work has followed this preliminary study. A study by Corvaglia et al. [180] determined that using NIR to measure the protein intake of preterm babies fed with breast milk was a suitable method to determine if the infant received the recommended amount. In a similar fashion, but at a point of care rather than at the bedside, Sauer and Kim [181] used NIR to analyze the milk of mother of preterm babies to adjust their food intake. Even though the number of sample was limited, authors reported very good correlation between measured and predicted fat, carbohydrate, and protein content for independent samples. No report of mother-to-mother variability was made.

Shaw et al. [182] performed some excellent analyses of urine samples. They used NIR to quantify protein, creatinine, and urea. Since water is not as big an interference in NIR as it is in mid-range IR, they easily carried out the analyses. Standard errors of prediction for the urea, creatinine, and protein were 16.6, 0.79, and 0.23 mM, respectively. They used 127 samples for each calibration. Both multiple linear regression (MLR) and PLS equations were generated, but PLS was eventually used to compensate for person-to-person variations. They concluded that the protein measurements would only be good for coarse screening, while the other two were comparable to current methods. The rapid nature of this test, using no reagents, is a marked improvement over the current clinically accepted method in terms of speed and throughput.

Further urine analyte analyses were performed by Jackson et al. and reported in 1997 [183]. Urine glucose, protein, urea, and creatinine concentrations were analyzed using rather simple algorithms. Urea, for instance, was calibrated by simply correlating with the absorbance at 2152 nm. Comparison with standard methods gave a linear relationship with a slope of nearly 1.00. Since creatinine and proteins are present in lower quantities and have lower absorptivities, a more complex algorithm, PLS, was needed to analyze the materials. The best correlation for creatinine delivered a slope of 0.953, and protein produced a slope of 0.923. In critical situations, where speed is more important than absolute numbers, NIR may be an important tool. Similar work was done by Pezzaniti et al., reporting in addition the possibility to measure ketone [184].

The measurement of glucose in urine has been a topic of significant research projects. Tanaka et al. [185] reported a prediction error of 22.3 mg/dl. Another study reported the same error rate of 20.6 mg/dl [186]. However, in a subsequent study, Tanaka et al. reported a decrease in the accuracy of the method to 60 mg/dl when several persons were used to develop the calibration model [187].

Two reports of estrus determination from urine measurements were reported by the team of Roumiana Tsenkova on the giant panda [188, 189]. The team reported good correlation between estrone-3-glucuronide and NIR spectra. No independent test set was available, however, to confirm the quantitative results. Using changes in the hydrogen-bonded water structures, authors reported that it is possible to predict ovulation with high accuracy by focusing the modeling on water bands.

The determination of biological exposure indexes was also tested with NIR. Hippuric acid is a urine component that is excreted in excess by individuals exposed to toluene. Ogawa et al. determined that the spectra has specific information related to hippuric acid and could be used for more frequent screenings to ensure workers' health [190].

As an example of ex vivo determinations, Shaw et al. were able to correlate the chemistry of synovial fluid, drawn from the knees of patients with arthritis diagnosis [191]. Conventional chemical analyses were performed on a series of patients with various types of arthritis; NIR was then used to analyze the fluids. A model equation was built using PLS. The identification of arthritis sufferers was remarkably good when the equation was tested on new patients.

Eye composition was examined in several interesting papers. The water content in bovine lenses was determined by Zink et al. [192]; Lopez-Gil et al. [193] measured the retinal image quality in the human eye as a function of the accommodation. The nerve fiber layer of an isolated rat retina was studied by Knighton and Huang [194].

Other blood constituents have also been measured. Blood lactate measurements were reported by Lafrance et al. with very high correlation compared with enzymatic measurements [195]. Kuenstner et al., in several papers, focused on the determination of hemoglobin levels [196–198]. Predictions of globulin, urea, albumin, and cholesterol have also been reported [199]. Note that in vivo lactate measurements were also published [200].

Disease states have also been investigated. Thus, the diagnosis of Parkinson's disease was shown to be possible by measuring oxidation stress by NIR [201]. Work was also performed on Alzheimer's disease by using the same oxidative stress measure [202].

While most in vivo measurements have been related to oxygen and glucose levels, the analysis of other fluids in vivo has been reported. In vivo pH measurements were reported by Zhang [203]. In this work, NIR was correlated with standard pH measurements to perform in vivo determination of the myocardial pH during regional ischemia. Dr. David H. Burns has spent a lot of research efforts in solving analytical issues to unlock the in vivo prediction of relevant biological components. At a recent conference, he presented work done on a new method to assess infection status in lactating mothers [204] and the general patient health during pregnancy by analyzing biofluids [205].

7.6   Fetuses and newborns

As stated by Liem and Greisen [206], "The most important cerebrovascular injuries in newborn infants, particularly in preterm infants, are cerebral haemorrhage and ischemic injury." Consequently, significant work has been done on the topic of oxygenation and newborns. Initial work was performed by Urlesberger et al. [207], who used NIR to analyze the behavior of cerebral oxygenation and blood volume in preterm infants during apnea. Park and Chang studied the effects of decreased cerebral perfusion, believing it induces cerebral ischemia and worsens brain damage in neonatal bacterial meningitis [208]. Numerous other workers have also studied the cranial blood of newborns and preterm babies [201–213]. Splanchnic oxygen delivery in neonates during apneicepisodes was studied by Petros et al. [214], while peripheral oxygen utilization in neonates was measured by Hassan et al. [215].

To not cover oxygenation all over again, readers are referred to the very good reviews written on the topic by Wolf et al. [216], Giliberti et al. [217], and specific infant studies [218–221]. Nevertheless, an interesting study by Aoyama et al. [222] looked at the difference in cerebral oxygenation when newborns were exposed to the odor of breast milk versus formula and found that breast milk provoked a significantly higher response.

Similar to the stool analysis used to diagnose organ deficiencies [162–167], Infante et al. used NIR to put in place a feeding regiment for constipated infants based on the measure of the stool's water content [223].

The analysis of amniotic fluid was used by Liu et al. to determine the fetal lung maturity [224], and Power et al. [225] looked at the possibility to assess preterm births. Authors found that NIR could differentiate between the metabolic profiles of second trimester women delivering at term and those at preterm.

The fetus was considered in a paper by Vishnoi et al. [226]. In this paper, they describe the measurement of photon migration through the fetal head in utero using a continuous wave NIR system. The same topic was addressed by Ramanujam et al. [227], using antepartum, transabdominal

NIR spectroscopy. A similar paper by Calvano et al. [228] discusses amni-oscopic endofetal illumination with infrared (NIR) guided fiber.

7.7 Cancer and precancer

Though in its infancy, NIR spectroscopy is finding applications in cancer research. The nonintrusive nature, as in blood chemistry work, is appeal-ing to any number of researchers. Two types of work have been con-ducted: modeling of cancer-related parameters and detection of tumors. The former relies on the use of chemometrics to develop prediction and classification of physiological elements pertaining to the presence or the development state of the tumor. The latter, on the other side, is focused on the detection of tumors based on the binding of probes or dyes that emit a light in the NIR region. A review of the use of NIR spectroscopy for cancer is presented here. The end of the section will focus on the use of imaging and dyes. For a review of the work done on skin cancer, refer to the tissue section (7.3).

Workers at Johns Hopkins University, under the tutelage of Chris Brown, worked on screening PAP smears using NIR spectroscopy [229]. Healthy patients, patients with abnormal cells, and patients with cervi-cal cancer were screened. Using discriminant analysis and principal component analysis, the samples were grouped and employed to exam-ine further samples. It was seen that malignant and healthy tissues were distinctly different, while abnormal tissues carried spectral features from both sets. New developments for the detection of cervical cancer by NIR have been published by Yang et al. [230]. The possibility to detect endome-trial cancer was also tested [231].

The diagnosis of colorectal cancer has been the focus of several stud-ies [232–234]. Researchers used the first and second overtone C–H stretch-ing to discriminate between cancer and normal tissue. The use of linear discriminant analysis, artificial neural networks, and clustering analy-sis was compared and very similar results obtained. While the former results were performed on resected samples, Shao et al. implemented an endoscope-based detection method to identify in vivo hyperplastic and adenomatous polyps [235]. Using a simple linear discriminant analysis based on principal component analysis scores, very good diagnostic sen-sitivities and specificities were obtained.

Mammograms are often uncomfortable and embarrassing for women. Using NIR, imaging [236] and an "optical biopsy" [237] may be performed. Since NIR radiation has some unique features, it has been suggested as an alternative to both x-rays and physical invasive biopsies. In younger women, with breasts dense to x-rays, NIR could be used to eliminate false positives. Since breast cancer is one of the leading causes of death and disfigurement in women, it is better detected early [236]. NIR has the advantage of not producing harmful radiation as in the x-ray approach, and women do not report as much discomfort. Using the possibility to predict hemoglobin by NIR, Pogue et al. [238] correlated the hemoglobin concentration levels with the presence of vascularity due to a tumor. Similar work was achieved by Gu et al. [239]. The oxygen consumption by tumors in breasts was also investigated [240]. NIR was used to qualify the effectiveness of therapy. By measuring the change in oxygen consumption of the tumor, researchers were able to evaluate how tumors were responding to treatments (i.e., photothermal therapy) [241, 242].

Magnetic resonance imaging (MRI) may be used in cases where x-rays are questionable, although MRI is not chemically specific and only shows masses more clearly than x-rays. Using NIR simultaneously could give a better picture of the mass's chemistry [243]. A time-resolved imager capable of acquiring images simultaneously was used for this work. Short-wavelength radiation in the range between 780 and 830 nm was seen to be best. Some important work was performed by Ntziachristos et al. [243], wherein they used both magnetic resonance imaging and NIR to afford precise co-registration of images and examined the potential and limitations of optical mammography. Using a time-resolved imager of their own design, the group acquired NIR images simultaneously with MR images. The intrinsic contrast at 780 and 830 nm was used to study the relative enhancement and kinetics due to the administration of infracyanamine R25, a NIR contrast agent. More recent work using both NIR and MRI was published by Saxena et al. [244] and Carpenter et al. [245].

Milne et al. [246] have developed stereotactically guided laser thermotherapy for breast cancer in situ measurements. They determine the temperature field within the breast to highlight potential tumors. A review paper by Tromberg et al. [247] discusses the noninvasive in vivo characterization of breast cancer tumors using photon migration spectroscopy. They compare the use of this technique with straightforward NIR spectroscopy. Near-infrared spectroscopic tomography is discussed in a paper by Tosteson et al. [248]. They extend basic concepts of statistical hypothesis testing and confidence intervals to images generated by this new procedure as used for breast cancer diagnosis. Tumor oxygenation was the topic of work published by Hull et al. [249]. In this paper, they showed detectable shifts toward higher saturation in all tumors. These hemoglobin changes were induced by introduction of carbogen. The new hemoglobin saturation levels were reached within a minute and remained constant throughout the experiment.

In 1994, Meurens et al. [250] determined that cryostat sections of carcinomatous tissue were different spectrally from noncarcinomatous tissue. Four distinct wavelength regions between 1200 and 2370 nm were found best for classification of the tissue samples. Samples included invasive ductal carcinoma, with a predominant intraductal component, mucinous carcinoma, and invasive lobular carcinoma. Despite the varied types of cancer cells, there was a distinct grouping of cancerous versus noncancerous cells. This work is being carried over to potential in vivo measurements.

Subcutaneous tumors were detected using NIR in a paper by Jarm et al. [251]. Low-level electric current used as electrotherapy on solid subcutaneous tumors was followed by NIR analysis of the tumor region. Data showed that oxygenation of the tumors was inhibited after the electrotherapy was applied.

A report on the detection of melanoma in the human eye was reported by Krohn et al. [252]. Authors reported that tumor tissues had a lower absorbance with weaker water bands and stronger absorption by hemoglobin.

While not directly related to NIR as we usually think of it, many researchers have taken advantage of the penetration power of NIR light to design dyes or probes whose location and quantity can be determined by imaging. A multitude of papers exist on the topic of creating molecular probes, activated by specific enzymes in tumors or targeted to particular receptors, overexpressed in tumors [253–258]. In many articles, imaging was used over single-point spectroscopy.

7.8 Photon migration in tissues

While some aspects of the theory of the separation of absorption and scattering and the related instrumentation have been discussed in Chapters 1 and 2, a significant amount of work has been done on applying these breakthroughs for studying photon migration in biological tissues. This section will summarize some of the work done on the topic.

An outstanding review by Steven L. Jacques summarizes reported tissue optical properties [259]. A majority of organs, body constituents, and tissues are sorted as a function of their scattering and absorption properties. The paper reports that skin elements are the tissues that scatter the light the most. Interestingly, the paper concludes that "the use of generic tissue can adequately mimic any real tissue," thus confirming the work performed by other researchers. The author also discusses the fact that there is so much variability between subjects, collection sites, and collection times that the actual optical properties should be estimated for individual patients rather than generalizing the findings to the population.

The first reports of absorption and scattering properties in tissues were studied on skin (1982 [260]), liver (1989 [261]), breast (1990 [262]), and skull (1993 [263]). In 1993, Arridge et al. published some work on a finite element approach for modeling photon transport in tissue [264]. In this method, called finite element modeling (FEM), the photon density in any object and the photon flux at its boundary allow modeling of light transport through tissue. Arridge and coworkers derived a mathematical model for one particular case. The calculation of the boundary flux is a function of time resulting from a δ-function point input to a two-dimensional circle (showing as a line source in an infinite cylinder) with homogeneous scattering and absorption properties. This model may be of some use to subsequent researchers, especially in NIR measurements of the human head, where scattering and light loss are extensive.

The effect of tissue thickness on signal intensity was studied by Lin et al. [265]. Using a modification of the random walk theory of photon migration, the team determined a strong dependence of the detected signal on tissue thickness up to approximately 2 mm. Light propagation in a semi-infinite layer of human tissue was carried out by Sadoghi [266]. Results showed that the optical discontinuity had a significant impact on the light distribution, thus making difficult an accurate estimation of tissue properties in resected tissues.

A multitude of papers have been published on mathematical models to determine tissue absorption and scattering properties [267–273]. It is, however, not the intent of this section to discuss how simulation algorithms were employed to determine the tissue parameters.

7.9   Review articles

The topic of functional near-infrared spectroscopy has been covered in detail by many reviews. Readers should refer to these papers to gain a better understanding of the work done on the topic. Below is a nonexhaustive list of reviews on the topic of NIR medical applications.

Blood glucose:
- Khalil, 1999 [5]
- Arnold and Small, 2005 [6]
- Vashist, 2012 [274]
- So et al., 2012 [275]

Oxygenation:
- Rooks et al., 2010 [70]
- Ferrari and Quaresima, 2012 [276]
- Soller et al., 2012 [277]
- Hampton and Schreiber, 2013 [278]

Major organs:
- Bhambhani, 2012 [279]
- Contini et al., 2012 [280]

Cancer:
- Mayevsky et al., 2003 [160]
- Tweddell et al., 2010 [175]
- Kravari et al., 2010 [176]

- Enfield and Gibson, 2012 [281]
- Pifferi et al., 2012 [282]
- Grosenick et al., 2012 [283]

Fetuses and newborns:

- Giliberti et al., 2011 [217]
- Wolf et al., 2012 [216]

Photon migration:

- Martelli, 2012 [284]
- Jacques, 2013 [259]

References

Blood glucose

1. F. M. Ham and G. M. Cohen, Non-Invasive Blood Glucose Monitoring, U.S. Patent 5553616 (1997).
2. G. Acosta, J. R. Henderson, A. N. Abul-Haj, T. L. Ruchti, S. L. Monfre, T. B. Blank, and K. H. Hazen, Compact Apparatus for Noninvasive Measurement of Glucose through Near-Infrared Spectroscopy, U.S. Patent 778787924 (2005).
3. G. Acosta, J. R. Henderson, A. N. Abul-Haj, T. L. Ruchti, S. L. Monfre, T. B. Blank, and K. H. Hazen, Method and Apparatus for Noninvasive Glucose Concentration Estimation through Near-Infrared Spectroscopy, U.S. Patent 0050107676 (2005).
4. J. W. Y. Chung, K. L. Fan, T. K. S. Wong, S. C. H. Lam, C. C. Cheung, C. M. Chan, and Y. K. Lau, Method for Predicting the Blood Glucose Level of a Person, U.S. Patent 7409239 (2008).
5. O. S. Khalil, Spectroscopic and Clinical Aspects of Noninvasive Glucose Measurements, *Clin. Chem.*, 45, 165 (1999).
6. M. A. Arnold and G. W. Small, Noninvasive Glucose Sensing, *Anal. Chem.*, 77, 5429 (2005).
7. M. Kohl, M. Cope, M. Essenpreis, and D. Boecker, Influence of Glucose Concentration on Light Scattering in Tissue Simulating Phantoms, *Opt. Lett.*, 19, 2170 (1994).
8. M. Kohl, M. Essenpreis, and M. Cope, Influence of Glucose Concentration upon the Transport of Light in Tissue-Simulating Phantoms, *Phys. Med. Biol.*, 40, 1267 (1995).
9. M. A. Arnold and L. A. Marquardt, Near-Infrared Spectroscopic Measurement of Glucose in a Protein Matrix, *Anal. Chem.*, 65, 3271 (1993).
10. M. A. Arnold, G. Small, and L. A. Marquart, Temperature-Insensitive Near-Infrared Spectroscopic Measurement of Glucose in Aqueous Solutions, *Appl. Spectrosc.*, 48(4), 477 (1994).
11. M. A. Arnold, S. Pan, H. Chung, and G. Small, Near-Infrared Spectroscopic Measurement of Physiological Glucose Levels in Variable Matrices of Protein and Triglycerides, *Anal. Chem.*, 68(7), 1124 (1996).
12. G. W. Small et al., Evaluation of Data Pretreatment and Model Building Methods for the Determination of Glucose from Near-Infrared Single-Beam Spectra, *Appl. Spectrosc.*, 53(4), 402 (1999).

13. J. J. Burmeister, M. A. Arnold, and G. W. Small, Spectroscopic Considerations for Noninvasive Blood Glucose Measurements with Near-Infrared Spectroscopy, *IEEE Lasers Electro-Opt. Soc.*, 12, 6 (1998).
14. M. R. Riley, M. A. Arnold, and D. W. Murhammer, Matrix-Enhanced Calibration Procedure for Multivariate Calibration Models with Near-Infrared Spectra, *Appl. Spectrosc.*, 52(10), 1339 (1998).
15. D. M. Haaland, M. R. Robinson, R. P. Eaton, and G. W. Koepp, Reagentless Near-Infrared Determination of Glucose in Whole Blood Using Multivariate Calibration, *Appl. Spectrosc.*, 46(10), 1575 (1992).
16. W. F. Schrader, Non-Invasive Anterior Chamber Glucose Monitoring by Near-Infrared Absorption Spectroscopy, an Alternative to Blood-Glucose Monitoring in Diabetic Patients? presented at Proceedings of 96th DOG Annual Meeting, 1998.
17. J. Backhaus et al., Device for the In Vivo Determination of an Optical Property of the Aqueous Humor of the Eye, U.S. Patent 5535743 (1996).
18. U. Fischer, K. Rebrin, T. Woedtke, and P. Able, Clinical Usefulness of the Glucose Concentration in the Subcutaneous Tissue—Properties and Pitfalls of Electrochemical Biosensors, *Horm. Metab. Res.*, 26, 515 (1994).
19. F. Sternberg, C. Meyerhoff, F. J. Memel, H. Meyer, F. Bischoff, and E. F. Pfeiffer, Subcutaneous Glucose Concentration: Its Real Estimation and Continuous Monitoring, *Diabetes Care*, 18, 1266 (1995).
20. M. A. Arnold, J. J. Burmeister, and H. Chung, Phantoms for Noninvasive Blood Glucose Sensing with Near-Infrared Transmission Spectroscopy, *Photochem. Photobiol.*, 67(1), 50 (1998).
21. M. A. Arnold, J. J. Burmeister, and G. Small, Phantom Glucose Calibration Models from Simulated Noninvasive Human Near-Infrared Spectra, *Anal. Chem.*, 70, 1773 (1998).
22. J. S. Maier et al., Possible Correlation between Blood Glucose Concentration and the Reduced Scattering Coefficient of Tissues in the Near-Infrared, *Opt. Lett.*, 19(24), 2026 (1994).
23. K. Jagemann et al., Applications of Near-Infrared Spectroscopy for Non-Invasive Determination of Blood/Tissue Glucose Using Neural Networks, *Z. Phys. Chem.*, 191, 179 (1995).
24. C. Fischbacher, U. A. Muller, B. Mentes, K. U. Jagermann, and K. Danzerk, Enhancing Calibration Models for Non-Invasive Near-Infrared Spectroscopic Blood Glucose Determination, *Fresenius J. Anal. Chem.*, 359, 78 (1997).
25. K. Danzer, U. A. Muller, B. Mertes, C. Fischbacker, and K. U. Jungemann, Near-Infrared Diffuse Reflection Spectroscopy for Non-Invasive Blood-Glucose Monitoring, *IEEE LEOS Newsl.*, 4(18), 1998.
26. H. Shamoon, I. Gabriely, R. Wozniak, M. Mevorach, J. Kaplan, and Y. Aharm, Transcutaneous Glucose Monitor during Hypoglycemia, presented at 5th Scientific Session, American Diabetes Association Meeting, San Diego, June 1999, paper 426.
27. K. Maruo, M. Tsurugi, M. Tamura, and Y. Ozaki, In Vivo Noninvasive Measurement of Blood Glucose by Near-Infrared Diffuse-Refectance Spectroscopy, *Appl. Spectrosc.*, 57, 1236 (2003).
28. H. M. Heise, R. Marboch, G. Janatsch, and J. D. Kruse-Jarres, Noninvasive Blood Glucose Assay by Near-Infrared Diffuse Reflectance Spectroscopy of the Human Inner Lip, *Appl. Spectrosc.*, 47(7), 875 (1993).

29. A. K. Amerov, J. Chen, G. W. Small, and M. A. Arnold, Scattering and Absorption Effects in the Determination of Glucose in Whole Blood by Near-Infrared Spectroscopy, *Anal. Chem.*, 77, 4587 (2005).

30. A. Lorber, Error Propagation and Figures of Merit for Quantification by Solving Matrix Equations, *Anal. Chem.*, 58, 1167 (1986).

31. M. Ren and M. A. Arnold, Comparison of Multivariate Calibration Models for Glucose, Urea, and Lactate from Near-Infrared and Raman Spectra, *Anal. Bioanal. Chem.*, 387, 879 (2007).

32. S. Kasemsumran, Y. P. Du, K. Maruo, and Y. Ozaki, Selective Removal of Interference Signals for Near-Infrared Spectra of Biomedical Samples by Using Region Orthogonal Signal Correction, *Anal. Chim. Acta*, 526, 193 (2004).

33. S. Kasemsumran, Y. P. Du, K. Maruo, and Y. Ozaki, Improvement of Partial Least-Squares Models for In Vitro and In Vivo Glucose Quantifications by Using Near-Infrared Spectroscopy and Searching Combination Moving Window Partial Least-Squares, *Chemometrics Intell. Lab. Syst.*, 82, 97 (2006).

34. L. N. Li, Q. B. Li, and G. J. Zhang, A Weak Signal Extraction Method for Human Blood Glucose Noninvasive Measurement Using Near Infrared Spectroscopy, *J. Infrared Milli. Terahz. Waves*, 30, 1191 (2009).

35. L. N. Li, Q. B. Li, and G. J. Zhang, A Nonlinear Model for Calibration of Blood Glucose Noninvasive Measurement Using Near Infrared Spectroscopy, *Infrared Physics Technol.*, 53, 410 (2010).

36. C. F. So, J. W. Y. Chung, M. S. M. Siu, and T. K. S. Wong, Improved Stability of Blood Glucose Measurement in Humans Using Near Infrared Spectroscopy, *Spectroscopy*, 25, 137 (2011).

37. K. Maruo, T. Oota, M. Tsurugi, T. Nakagawa, H. Arimoto, M. Tamura, Y. Ozaki, and Y. Yamada, New Methodology to Obtain a Calibration Model for Noninvasive Near-Infrared Blood Glucose Monitoring, *Appl. Spectrosc.*, 60, 441 (2006).

Blood oxygenation

38. F. F. Jobsis, Noninvasive, Infrared Monitoring of Cerebral and Myocardial Oxygen Sufficiency and Circulatory Parameters, *Science*, 198, 1264 (1977).

39. Y. Ozaki, A. Mizuno, T. Hayashi, K. Tashibu, S. Maraishi, and K. Kawauchi, Nondestructive and Noninvasive Monitoring of Deoxyhemoglobin in the Vein by Use of a Near-Infrared Reflectance Spectrometer with a Fiber-Optic Probe, *Appl. Spectrosc.*, 46(1), 180 (1992).

40. M. G. Sowa et al., Noninvasive Assessment of Regional and Temporal Variations in Tissue Oxygenation by Near-Infrared Spectroscopy and Imaging, *Appl. Spectrosc.*, 51(2), 143 (1997).

41. D. M. Mancini, L. Bolinger, H. Li, K. Kendrick, B. Chance, and J. R. Wilson, Validation of Near-Infrared Spectroscopy in Humans, *J. Appl. Physiol.*, 77(6), 2740 (1994).

42. L. Lin and D. A. York, Two-Layered Phantom Experiments for Characterizing the Influence of a Fat Layer on Measurement of Muscle Oxygenation Using NIRS, *Proc. SPIE*, 3257 (1998).

43. K. Yamamoto et al. Near-Infrared Muscle Oximeter That Can Correct the Influence of a Subcutaneous Fat Layer, *Proc. SPIE*, 3257 (1998).

44. M. A. Franceschini et al., *The Effect of Water in the Quantitation of Hemoglobin Concentration in a Tissue-Like Phantom by Near-Infrared Spectroscopy*, Optical Society of America, Washington, DC, 1996.
45. Z. X. Jiang et al., Novel NIR Instrument for Non-Invasive Monitoring and Quantification of Cerebral Tissue Oxygenation, *Proc. SPIE*, 3257 (1998).
46. R. Marbach and H. M. Heise, Optical Diffuse Reflectance Accessory for Measurements of Skin Tissue by Near-Infrared Spectroscopy, *Appl. Opt.*, 34(4), 610 (1995).
47. K. Miyasaka, NIRS Use in Pediatric Anesthesia and ICU, presented at 96 PICU Conference, Toronto, 1996.
48. S. Van Huffel and P. Casaer, Changes in Oxygenation and Hemodynamics in Neonatal Brain by Means of Near-Infrared Spectroscopy: A Signal Analysis Study, 1997, http://www.esat.kuleuven.be/sista/yearreport96/node6.html (accessed January 18, 2014).
49. S. Van Huffel, P. Casaer, and M. Kessler, Modeling and Quantification of Chromophore Concentrations, Based on Optical Measurements in Living Tissues, 1998, http://www.esat.kuleuven.be/sista/yearreport97/node33.html (accessed January 18, 2014).
50. C. E. Cooper, J. Torres, M. Sharpe, and M. T. Wilson, The Relationship of Oxygen Delivery to Absolute Hemoglobin Oxygenation and Mitochondrial Cytochrome Oxidase Redox State in the Adult Brain: A Near-Infrared Spectroscopy Study, *Biochem. J.*, 332, 627 (1998).
51. J. S. Wyatt, M. Cope, D. T. Delp, C. E. Richardson, A. D. Edwards, S. Wray, and E. O. Reynolds, Quantitation of Cerebral Blood Volume in Human Infants by Near-Infrared Spectroscopy, *J. Appl. Physiol.*, 68, 1086 (1990).
52. V. V. Kupriyanov, R. A. Shaw, B. Xiang, H. Mantsch, and R. Deslauriers, Oxygen Regulation of Energy Metabolism in Isolated Pig Hearts: A Near-IR Spectroscopy Study, *J. Mol. Cell. Cardiol.*, 29, 2431 (1997).
53. J. R. Mansfield et al., Fuzzy C-Means Clustering and Principal Component Analysis of Time Series from Near-Infrared Imaging of Forearm Ischemia, *Comput. Med. Imaging Graphics*, 21(5), 299 (1997).
54. T. Wolf, U. Lindauer, H. Obrig, J. Drier, J. Back, A. Villringer, and O. Dirnagl, *J. Cereb. Blood Flow Metab.*, 16, 1100 (1996).
55. K. D. Liem, J. C. Hopman, L. A. Kollee, and R. Oeseburg, Effects of Repeated Indomethacin Administration on Cerebral Oxygenation and Hemodynamics in Pre-Term Infants: Combined Near-Infrared Spectrophotometry and Doppler Ultrasound Study, *Eur. J. Pediatr.*, 153(7), 504 (1994).
56. T. J. Germon, P. D. Evans, A. R. Manara, N. J. Barnett, P. Wall, and R. J. Nelson, Sensitivity of Near Infrared Spectroscopy to Cerebral and Extra-Cerebral Oxygen Changes Is Determined by Emitter-Detector Separation, *J. Clin. Monit.*, 10, 1 (1998).
57. L. C. Henson, C. Calalang, J. A. Temp, and D. S. Ward, Accuracy of a Cerebral Oximeter in Healthy Volunteers under Conditions of Isocapnic Hypoxia, *Anesthesiology*, 88(1), 58 (1998).
58. Y. Hoshi, H. Onoe, Y. Watanabe, J. Andersson, M. Bergstram, A. Lilja, B. Langstrom, and M. Tamura, Non-Synchronous Behavior of Neuronal Activity, Oxidative Metabolism, and Blood Supply during Mental Tasks in Man, *Neurosci. Lett.*, 172, 129 (1994).

59. F. Okada, Y. Tokumitsu, Y. Hoshi, and M. Tamura, Impaired Interhemispheric Integration in Brain Oxygenation and Hemodynamics in Schizophrenia, *Euro. Arch. Pschiatry Clin. Neurosci.*, 244, 17 (1994).

60. Y. Hoshi, S. Mizukami, and M. Tamura, Dynamic Features of Hemodynamic and Metabolic Changes in the Human Brain during All-Night Sleep as Revealed by Near-Infrared Spectroscopy, *Brain Res.*, 652, 257 (1994).

61. A. Akına, D. Bilensoya, U. E. Emira, M. Gülsoya, S. Candansayarc, and H. Bolay, Cerebrovascular Dynamics in Patients with Migraine: Near-Infrared Spectroscopy Study, *Neurosci. Lett.*, 400, 86 (2006).

62. G. Chow et al., The Relation between Pump Flow Rate and Pulsatility on Cerebral Hemodynamics during Pediatric Cardiopulmonary Bypass, *J. Thorac. Cardiovasc. Surg.*, 114(4), 1123 (1997).

63. R. Totaro, G. Barattelli, V. Quaresima, A. Carolei, and M. Ferrari, Evaluation of Potential Factors Affecting the Measurement of Cerebrovascular Reactivity by Near-Infrared Spectroscopy, *Clin. Sci.*, 95, 497 (1998).

64. Hitachi, Team Develops World's First Light-Based Procedure for Examining Epileptic Brain Sites, http://www.hitachi.com/New/cnews/E/1997/970710B.html (accessed January 18, 2014).

65. D. A. Boas, T. Gaudette, G. Strangman, X. Cheng, J. J. A. Marota, and J. B. Mandeville, The Accuracy of Near Infrared Spectroscopy and Imaging during Focal Changes in Cerebral Hemodynamics, *Neuroimage* 13, 76 (2001).

66. G. Strangman, M. A. Franceschini, and D. A. Boas, Factors Affecting the Accuracy of Near-Infrared Spectroscopy Concentration Calculations for Focal Changes in Oxygenation Parameters, *Neuroimage*, 18, 865 (2003).

67. K. Ide, A. Horn, and N. H. Secher, Cerebral Metabolic Response to Submaximal Exercise, *J. Appl. Physiol.*, 87, 1604 (1999).

68. Y. Bhambhani, R. Malik, and S. Mookerjee, Cerebral Oxygenation Declines at Exercise Intensities above the Respiratory Compensation Threshold, *Respir. Physiol. Neurobiol.*, 156, 196 (2007).

69. A. Timinkul, M. Kato, T. Omori, C. C. Deocaris, A. Ito, T. Kizuka, Y. Sakairi, T. Nishijima, T. Asada, and H. Soya, Enhancing Effect of Cerebral Blood Volume by Mild Exercise in Healthy Young Men: A Near-Infrared Spectroscopy Study, *Neurosci. Res.*, 61, 242 (2008).

70. C. R. Rooks, N. J. Thom, K. K. McCully, and R. K. Dishman, Effects of Incremental Exercise on Cerebral Oxygenation Measured by Near-Infrared Spectroscopy: A Systematic Review, *Progr. Neurobiol.*, 92, 134 (2010).

71. M. J. Herrmann, A. C. Ehlis, and A. J. Fallgatter, Frontal Activation during a Verbal-Fluency Task as Measured by Near-Infrared Spectroscopy, *Brain Res. Bull.*, 61, 51 (2003).

72. P.-H. Chou and T.-H. Lan, The Role of Near-Infrared Spectroscopy in Alzheimer;s Disease, *J. Clin. Gerontol. Geriatr.*, 4(2), 33 (2013).

73. C. Pennekamp, M. Bots, L. Kappelle, et al., The Value of Near-Infrared Spectroscopy Measured Cerebral Oximetry during Carotid Endarterectomy in Perioperative Stroke Prevention: A Review, *Eur. J. Vasc. Endovasc. Surg.*, 38, 539 (2009).

74. J. K.-J. Li, T. Wang, and H. Zhang, Rapid Noninvasive Continuous Monitoring of Oxygenation in Cerebral Ischemia and Hypoxia, *Cardiovasc. Eng.*, 10, 213 (2010).

75. C. Robertson, S. Gopinath, and B. Chance, A New Application for Near-Infrared Spectroscopy: Detection of Delayed Intracranial Hematomas after Head Injury, *J. Neurotrauma*, 12, 591 (1995).
76. A. Gill, K. Rajneesh, C. Owen, et al., Early Optical Detection of Cerebral Edema In Vivo, *J. Neurosurg.* 114, 470 (2011).
77. H. Vohra, A. Modi, and S. Ohri, Does Use of Intra-Operative Cerebral Regional Oxygen Saturation Monitoring during Cardiac Surgery Lead to Improved Clinical Outcomes? *Interact. Cardiovasc. Thorac. Surg.*, 9, 318 (2009).
78. J. Murkin, NIRS: A Standard of Care for CPB vs. an Evolving Standard for Selective Cerebral Perfusion? *J. Extra Corpor. Technol.*, 41, 11 (2009).
79. D. Highton, C. Elwell, and M. Smith, Noninvasive Cerebral Oximetry: Is There Light at the End of the Tunnel? *Curr. Opin. Anaesthesiol.*, 23(5), 576 (2010).
80. J. F. Jabre, G. M. Squintani, and K. K. H. Chui, Oxyneurography: A New Technique for the Measurement of Nerve Oxygenation, *Muscle Nerve*, 45, 75 (2012).
81. A. A. Oraevsky, E. V. Savateeva, S. V. Solomatin, A. Karabutov, V. G. Andreev, Z. Gatalica, T. Khamapirad, and P. M.Henrichs, Optoacoustic Imaging of Blood for Visualization and Diagnostics of Breast Cancer *Proc. SPIE*, 4618, 81 (2002).
82. R. G. M. Kolkman, E. Hondebrink, W. Steenbergen, and F. F. M. de Mul, In Vivo Photoacoustic Imaging of Blood Vessels Using an Extreme-Narrow Aperture Sensor, *IEEE J. Sel. Top. Quantum Electron.*, 9, 343 (2003).
83. J. J. Niederhauser, M. Jaeger, R. Lemor, P. Weber, and M. Frenz, Combined Ultrasound and Optoacoustic System for Real Time High Contrast Vascular Imaging In Vivo, *IEEE Trans. Med. Imaging*, 24, 436 (2005).
84. E. Z. Zhang and P. C. Beard, 2D Backward-Mode Photoacoustic Imaging System for NIR (650–1200 nm) Spectroscopic Biomedical Applications, *Proc. SPIE* 6086 (2006).
85. J. Laufer, D. Delpy, C. Elwell, and P. Beard, Quantitative Spatially Resolved Measurement of Tissue Chromophore Concentrations Using Photoacoustic Spectroscopy: Application to the Measurement of Blood Oxygenation and Haemoglobin Concentration, *Phys. Med. Biol.*, 52, 141 (2007).
86. R. Boushel, F. Pott, P. Madsen, G. Radegran, M. Nowak, B. Quistroff, and N. Secher, Muscle Metabolism from Near-Infrared Spectroscopy during Rhythmic Handgrip in Humans, *Eur. J. Appl. Physiol. Occup. Physiol.*, 79(1), 41 (1998).
87. A. Lassnigg, M. Hiemayr, P. Keznicki, T. Mullner, M. Ehrlich, and G. Grubhofer, Cerebral Oxygenation during Cardiopulmonary Bypass Measured by Near-Infrared Spectroscopy: Effects of Hemodilution, Temperature, and Flow, *J. Cardiothorac. Vasc. Anesth.*, 13(5), 544 (1999).
88. A. T. Lovell, H. Owen-Reece, C. E. Elwell, M. Smith, and J. C. Goldstone, Continuous Measurement of Cerebral Oxygenation by Near-Infrared Spectroscopy during Induction of Anesthesia, *Anesth. Analg.*, 88(3), 554 (1999).
89. T. Higami, S. Kozawa, T. Asada, H. Obo, K. Gan, K. Iwahashi, and H. Nohara, Retrograde Cerebral Perfusion versus Selective Cerebral Perfusion as Evaluated by Cerebral Oxygen Saturation during Aortic Arch Reconstruction, *Ann. Thorac. Surg.*, 67(4), 1091 (1999).

90. G. Liu, I. Burcev, F. Pott, K. Ide, I. Horn, and N. H. Sander, Middle Cerebral Artery Flow Velocity and Cerebral Oxygenation during Abdominal Aortic Surgery, *Anaesth. Intensive Care*, 27(2), 148 (1999).

91. M. L. Blas, E. B. Lobato, and T. Martin, Non-Invasive Infrared Spectroscopy as a Monitor of Retrograde Cerebral Perfusion during Deep Hypothermia, *J. Cardiothorac. Vasc. Anesth.*, 13(2), 244 (1999).

92. M. J. Van de Ven, W. N. Colier, B. T. Kersten, B. Oeseburg, and H. Folgering, Cerebral Blood Volume Responses to Acute $PaCO_2$ Changes in Humans, Assessed with Near-Infrared Spectroscopy, *Adv. Exp. Med. Biol.*, 471, 199 (1999).

93. M. Wolf, O. Weber, M. Keel, X. Golay, M. Scheidegger, H. U. Bucher, S. Kollias, P. Boesiger, and O. Banziger, Comparison of Cerebral Blood Volume Measured by Near-Infrared Spectroscopy and Contrast Enhanced Magnetic Resonance Imaging, *Adv. Exp. Med. Biol.*, 471, 767 (1999).

94. J. Kytta, J. Ohman, P. Tanskanen, and T. Randell, Extracranial Contribution to Cerebral Oximetry in Brain Dead Patients: A Report of Six Cases, *J. Neurosurg. Anesthesiol.*, 11(4), 252 (1999).

95. C. Casavola, L. A. Paunescu, S. Fantini, M. A. Franceshini, P. M. Lugara, and E. Gratton, Application of Near-Infrared Tissue Oximetry to the Diagnosis of Peripheral Vascular Disease, *Clin. Hemorheol. Microcirc.*, 21(3–4), 389 (1999).

96. W. N. Colier, V. Quaresima, B. Oeseburg, and M. Ferrari, Human Motor-Cortex Oxygenation Changes Induced by Cyclic Coupled Movements of Hand and Foot, *Exp. Brain Res.*, 129(3), 457 (1999).

97. K. Krakow, S. Ries, M. Daffertshofer, and M. Hennerici, Simultaneous Assessment of Brain Tissue Oxygenation and Cerebral Perfusion during Orthostatic Stress, *Eur. Neurol.*, 43(1), 39 (2000).

98. G. Fuchs, G. Schwartz, A. Kulier, and G. Litscher, The Influence of Positioning on Spectroscopic Measurements of Brain Oxygenation, *J. Neurosurg. Anesthesiol.*, 12(2), 75 (2000).

99. A. J. Spielman, G. Zhang, C. M. Yang, S. Serizawa, M. Nagata, H. von Gizycki, and R. R. Alfano, Intercerebral Hemodynamics Probed by Near-Infrared Spectroscopy in the Transition between Wakefulness and Sleep, *Brain Res.*, 866(1–2), 313 (2000).

100. Y. Nagashima, Y. Yada, M. Hattori, and A. Sakai, Development of a New Instrument to Measure Oxygen Saturation and Total Hemoglobin Volume in Local Skin by Near-Infrared Spectroscopy and Its Clinical Application, *Int. J. Biometrol.*, 44(1), 9 (2000).

101. B. A. McKinley, R. G. Marvin, Concanour, and F. A. Moore, Tissue Hemoglobin O_2 Saturation during Resuscitation of Traumatic Shock Monitored Using Near-Infrared Spectrometry, *J. Trauma-Injury Infection Crit. Care*, 48(4), 637 (2000).

102. R. Kragelj, T. Jarm, and D. Miklavcic, Reproducibility of Parameters of Postocclusive Reactive Hyperemia Measured by Near-Infrared Spectroscopy and Transcutaneous Oximetry, *Ann. Biomed. Eng.*, 28(2), 168 (2000).

103. R. Boushel, H. Langberg, J. Olesen, J. Gonzales-Alonzo, J. Bülow, and M. Kjær, Monitoring Tissue Oxygen Availability with Near Infrared Spectroscopy (NIRS) in Health and Disease, *Scand. J. Med. Sci. Sports*, 11(4), 213 (2001).

104. S. Goldman, F. Sutter, F. Ferdinand, and C. Trace, Optimizing Intraoperative Cerebral Oxygen Delivery Using Noninvasive Cerebral Oximetry Decreases the Incidence of Stroke for Cardiac Surgical Patients, *Heart Surg. Forum*, 7(5), 392 (2004).

105. G. M. Hoffman, Pro: Near-Infrared Spectroscopy Should Be Used for All Cardiopulmonary Bypass, *J. Cardiothorac. Vasc. Anesth.*, 20(4), 606 (2006).

106. M. Wolf, M. Ferrari, and V. Quaresima, Progress of Near-Infrared Spectroscopy and Topography for Brain and Muscle Clinical Applications, *J. Biomed. Opt.*, 12(6) (2007).

107. M. Wolf, G. Morren, D. Haensse, T. Karen, U. Wolf, J. C. Fauchère, and H. U. Bucher, Near Infrared Spectroscopy to Study the Brain: An Overview, *Optoelectron. Rev.*, 16(4), 413 (2008).

108. M. Schecklmann, A. C. Ehlis, M. M. Plichta, and A. J. Fallgatter, Influence of Muscle Activity on Brain Oxygenation during Verbal Fluency Assessed with Functional Near-Infrared Spectroscopy, *Neuroscience*, 171, 434 (2010).

109. J.-F. Georger, O. Hamzaoui, A. Chaari, J. Maizel, C. Richard, and J.-L. Teboul, Restoring Arterial Pressure with Norepinephrine Improves Muscle Tissue Oxygenation Assessed by Near-Infrared Spectroscopy in Severely Hypotensive Septic Patients, *Intensive Care Med.*, 36, 1882 (2010).

110. Y. Shang, R. Cheng, L. Dong, S. J. Ryan, S. P. Saha, and G. Yu, Cerebral Monitoring during Carotid Endarterectomy Using Near-Infrared Diffuse Optical Spectroscopies and Electroencephalogram, *Phys. Med. Biol.*, 56, 3015 (2011).

111. M. F. Oliveira, M. K. Rodrigues, E. Treptow, T. M. Cunha, E. M. V. Ferreira, and J. A. Neder, Effects of Oxygen Supplementation on Cerebral Oxygenation during Exercise in Chronic Obstructive Pulmonary Disease Patients Not Entitled to Long-Term Oxygen Therapy, *Clin. Physiol. Funct. Imaging*, 32, 52 (2012).

112. S. M. Bailey, K. D. Hendricks-Muñoz, and P. Mally, Splanchnic-Cerebral Oxygenation Ratio as a Marker of Preterm Infant Blood Transfusion Needs, *Transfusion*, 52(2), 252 (2012).

Tissue

113. E. Dreassi, G. Ceramelli, L. Fabbri, F. Vocioni, P. Bartalini, and P. Corti, Application of Near-Infrared Reflectance Spectroscopy in the Study of Atopy. Part 1. Investigation of Skin Spectra, *Analyst*, 122(8), 767 (1997).

114. E. Dreassi, G. Ceramelli, P. Bura, P. L. Perruccio, F. Vocioni, P. Bartalini, and P. Corti, Application of Near-Infrared Reflectance Spectroscopy in the Study of Atopy. Part 2. Interactions between the Skin and Polyethylene Glycol 400, Isopropyl Myristate, and Hydrogel, *Analyst*, 122(8), 771 (1997).

115. P. Corti, G. Ceramelli, E. Dreassi, and S. Mattii, Application of Near-Infrared Reflectance Spectroscopy in the Study of Atopy. Part 3. Interactions between the Skin and Fomblins, *Analyst*, 122(8), 788 (1997).

116. M. F. Stranc, M. G. Sowa, B. Abdulrauf, and H. H. Mentsch, Assessment of Tissue Viability Using Near-Infrared Spectroscopy, *Br. J. Plast. Surg.*, 51, 210 (1998).

117. M. Attas, M. Hewko, J. Payette, T. Posthumus, M. Sowa, and H. Mantsch, Visualization of Cutaneous Hemoglobin Oxygenation and Skin Hydration Using Near-Infrared Spectroscopic Imaging, *Skin Res. Technol.*, 7, 238 (2001).

118. A. Repez, D. Oroszy, and Z. M. Arnez, Continuous Postoperative Monitoring of Cutaneous Free Flaps Using Near Infrared Spectroscopy, *J. Plast. Reconstr. Aesthet. Surg.*, 61, 71e77 (2008).

119. K. M. Cross, L. Leonardi, J. R. Payette, M. Gomez, M. A. Levasseur, B. J. Schattka, M. G. Sowa, and J. S. Fish, Clinical Utilization of Near-Infrared Spectroscopy Devices for Burn Depth Assessment, *Wound Rep. Reg.*, 15, 332 (2007).

120. K. M. Cross, L. Leonardi, M. Gomez, J. R. Freisen, M. A. Levasseur, B. J. Schattka, M. G. Sowa, and J. S. Fish, Noninvasive Measurement of Edema in Partial Thickness Burn Wounds, *J. Burn Care Res.*, 30(5), 807 (2009).

121. N. E. I. Langlois, The Science Behind the Quest to Determine the Age of Bruises: A Review of the English Language Literature, *Forens. Sci. Med. Pathol.*, 3, 241 (2007).

122. M. G. Sowa, J. Wang, C. P. Schultz, M. K. Ahmed, and H. H. Mantsch, Infrared Spectroscopic Investigation of In Vivo and Ex Vivo Human Nails, *Vib. Spectrosc.*, 10, 49 (1995).

123. N. Kasa and K. M. Heinonen, Near-Infrared Interactance in Assessing Superficial Body Fat in Exclusively Breastfed, Full-Term Neonates, *Acta Paediatr.*, 82, 1 (1993).

124. K. Z. Liu, T. C. Dembinski, and H. H. Mantsch, Prediction of Fetal Lung Maturity from Near-Infrared Spectra of Amniotic Fluid, *Int. J. Gynecol. Obstet.*, 57, 161 (1997).

125. C. H. Barlow, K. A. Kelly, and J. J. Kelly, Tissue Temperature by Near-Infrared Spectroscopy, in Optical Tomography, Photon Migration, and Spectroscopy of Tissue and Model Media, *Proc. SPIE*, 2389, 818 (1995).

126. C. L. Tsai, J. C. Chen, and W. J. Wang, Absorption Properties of Soft Tissue Constituents in 900–1340 nm Region, *Proc. SPIE*, 3257 (1998).

127. J. M. Schmitt, H. Yang, and J. N. Qu, Interpretation and Processing of NIR Spectra of Turbid Biological Tissue, *Proc. SPIE*, 3257 (1998).

128. S. Andersson-Engels, R. Berg, S. Svanberg, and O. Jarlman, Time-Resolved Transillumination of Tissue for Medical Diagnostics, *Opt. Lett.*, 15(21), 1179 (1990).

129. R. Berg et al., Time-Resolved Transillumination Imaging, in *Medical Optical Tomography: Functional Imaging and Monitoring*, vol. 11, ed. G. Muller et al., SPIE Institute, Bellingham, WA, 1993, p. 397.

130. O. Jarlman, R. Berg, and S. Svanberg, Time-Resolved Transillumination of the Breast, *Acta Radiol.*, 33, 277 (1992).

131. R. Berg, S. Andersson-Engels, and K. Rama, Medical Transillumination Imaging Using Short Pulse Diode Lasers, *Appl. Opt.*, 32, 574 (1993).

132. S. Andersson-Engels, R. Berg, and K. Rama, Time-Resolved Transillumination of Tissue and Tissue-Like Phantoms for Medical Diagnostics, *Proc. SPIE*, 2081, 137 (1993).

133. L. M. McIntosh, R. Summers, M. Jackson, H. H. Mantsch, J. R. Mansfield, M. Howlett, A. Neil Crowson, and J. W. P. Toole, Towards Non-Invasive Screening of Skin Lesions by Near-Infrared Spectroscopy, *J. Invest. Dermatol.*, 116, 175 (2001).

134. E. Salomatina, B. Jiang, J. Novak, and A. N. Yaroslavsky, Optical Properties of Normal and Cancerous Human Skin in the Visible and Near-Infrared Spectral Range, *J. Biomed. Opt.*, 11 (2006).

135. N. Neittaanmäki-Perttu, M. Gronroos, T. Tani, I. Polonen, A. Ranki, O. Saksela, and E. Snellman, Detecting Field Cancerization Using a Hyperspectral Imaging System, *Lasers Surg. Med.*, 45, 410 (2013).

136. R. A. Lodder and L. Cassis, Arterial Analysis with a Novel Near-IR Fiber-Optic Probe, *Spectroscopy*, 5(7), 12 (1990).

137. L. A. Cassis and R. A. Lodder, Near-IR Imaging of Atheromas in Living Arterial Tissue, *Anal. Chem.*, 65, 1247 (1993).

138. W. Jaross, V. Neumeister, P. Lattke, et al., Determination of Cholesterol in Atherosclerotic Plaques Using Near Infrared Diffuse Reflection Spectroscopy, *Atherosclerosis*, 147, 327 (1999).

139. V. Neumeister, M. Scheibe, P. Lattke, and W. Jaross, Determination of the Cholesterol-Collagen Ratio of Arterial Atherosclerotic Plaques Using Near Infrared Spectroscopy as a Possible Measure of Plaque Stability, *Atherosclerosis*, 165, 251 (2002).

140. J. Wang, Y. J. Geng, B. Guo, et al., Near-Infrared Spectroscopic Characterization of Human Advanced Atherosclerotic Plaques, *J. Am. Coll. Cardiol.*, 39, 1305 (2002).

141. P. Moreno, R. Lodder, K. Purushothaman, et al., Detection of Lipid Pool, Thin Fibrous Cap, and Inflammatory Cells in Human Aortic Atherosclerotic Plaques by Near-Infrared Spectroscopy, *Circulation*, 105, 923 (2002).

142. P. R. Moreno, S. E. Ryan, et al., Identification of Lipid-Rich Plaques in Human Coronary Artery Autopsy Specimens by Near-Infrared Spectroscopy, *J. Am. Coll. Cardiol.*, 37(Suppl 2), A356 (2002).

143. J. D. Caplan, S. Waxman, R. W. Nesto, and J. E. Muller, Near-Infrared Spectroscopy for the Detection of Vulnerable Coronary Artery Plaques, *J. Am. Coll. Cardiol.*, 47(8, Suppl C), C92 (2006).

144. S. Waxman, J. Tang, et al., In Vivo Detection of a Coronary Artificial Target with a Near Infrared Spectroscopy Catheter, *Am. J. Cardiol.*, 94(Suppl 6A), 141E (2004).

145. S. Waxman, Near-Infrared Spectroscopy for Plaque Characterization, *J. Interventional Cardiol.*, 21(6), 452 (2008).

146. K. Fall, A. Maehara, and G. S. Mintz, Intravascular Imaging in Patients with Acute Coronary Syndromes and Unstable Coronary Plaques, *Curr. Cardiovasc. Imaging Rep.*, 4(4), 269 (2011).

147. A. M. Fard, P. Vacas-Jacques, E. Hamidi, H. Wang, R. W. Carruth, J. A. Gardecki, and G. J. Tearney, Optical Coherence Tomography—Near Infrared Spectroscopy System and Catheter for Intravascular Imaging, *Opt. Express*, 21(25), 30849 (2013).

148. K. Jansen, M. Wu, A. F. W. Van der Steen, and G. Van Soest, Photoacoustic Imaging of Human Coronary Atherosclerosis in Two Spectral Bands, *Photoacoustics*, 2(1), 12 (2014).

149. J. M. G. Canvin, S. Bernatsky, C. A. Hitchon, M. Jackson, M. G. Sowa, J. R. Mansfield, H. H. Eysel, H. H. Mantsch, and H. S. El-Gabalawy, Infrared Spectroscopy: Shedding Light on Synovitis in Patients with Rheumatoid Arthritis, *Rheumatology* 42, 76 (2003).

Major organs

150. T. Sato, Yasanuma, T. Kusano, N. Sasaki, Y. Shindo, and K. Koyama, Difference in Hepatic Issue Oxygenation between Total Vascular Exclusion and Inflow Occlusion of the Liver and the Possible Role of Hepatic Venous Blood under Liver Ischemia, *Dig. Surg.*, 15(1), 15 (1998).

151. L. R. Jiao, A. M. Seifalian, N. Habib, and B. R. Davidson, The Effect of Mechanically Enhancing Portal Venous Inflow on Hepatic Oxygenation, Microcirculation, and Function in a Rabbit Model with Extensive Hepatic Fibrosis, *Hepatology*, 30(1), 46 (1999).

152. L. R. Jiao, A. A. El-Desoky, A. M. Seifalian, N. Habib, and B. R. Davidson, Effect of Liver Blood Flow and Function on Hepatic Indocyanine Green Clearance Measured Directly in a Cirrhotic Animal Model, *Br. J. Surg.*, 87(5), 568 (2000).

153. A. E. El-Dosoky, A. Seifalian, M. Cope, D. Delphy, and B. Davidson, Changes in Tissue Oxygenation of the Porcine Liver Measured by Near-Infrared Spectroscopy, *Liver Transpl. Surg.*, 5(3), 219 (1999).

154. J. Teller, K. Schwendener, M. Wolf, M. Keel, H. U. Bucher, S. Fanconi, and O. Baenziger, Continuous Monitoring of Liver Oxygenation with Near Infrared Spectroscopy during Nasogastric Tube Feeding in Neonates, *J. Suisse Med.*, 130(18), 652 (2000).

155. C. T. Germer, A. Roggan, J. P. Ritz, C. Isbert, D. Albrecht, G. Muller, and H. J. Buhr, Optical Properties of Native and Coagulated Human Liver Tissue and Liver Metastases in the Near-Infrared Range, *Lasers Surg. Med.*, 23(4), 194 (1998).

156. G. Schulz, M. Weiss, U. Bauersfeld, J. Teller, D. Haensse, H. Bucher, and O. Baenziger, Liver Tissue Oxygenation as Measured by Near-Infrared Spectroscopy in the Critically Ill Child in Correlation with Central Venous Oxygen Saturation, *Intensive Care Med.*, 28(2), 184 (2002).

157. J. Cochrane, B. T. Williams, A. Banerjee, A. H. Harken, T. J. Burle, C. B. Cairns, and J. I. Shapiro, Ischemic Preconditioning Attenuates Functional, Metabolic, and Morphologic Injury from Ischemic Acute Renal Failure in the Rat, *Renal Failure*, 21(2), 135 (1999).

158. N. J. Lane, M. S. Thorniley, S. Manek, B. J. Fuller, and C. J. Green, Hemoglobin Oxygenation Kinetics and Secondary Ischemia in Renal Transplantation, *Transplantation*, 61(5), 689 (1996).

159. E. Vidal, A. Amigoni, V. Brugnolaro, G. Ghirardo, P. Gamba, A. Pettenazzo, G. F. Zanon, C. Cosma, M. Plebani, and L. Murer, Near-Infrared Spectroscopy as Continuous Real-Time Monitoring for Kidney Graft Perfusion, *Pediatr. Nephrol.*, 29(5), 909 (2014).

160. A. Mayevsky, J. Sonn, M. Luger-Hamer, and R. Nakache, Real-Time Assessment of Organ Vitality during the Transplantation Procedure, *Transplantation Rev.*, 17(2), 96 (2003).

161. D. Choudhury, B. Michener, P. Fennelly, and M. Levi, Near-Infrared Spectroscopy in the Evaluation of Lower Limb Ischemia in Hemodialysis Patients, *J. Vasc. Technol.*, 23(1), 21 (1999).

162. P. Faggionto, V. Stanzial, A. Facchin, R. Rolfini, L. Zanolla, G. Cabrini, and G. Mastella, Near-Infrared Reflectance Analysis for the Evaluation of the Exocrine Pancreatic Function, *Eur. J. Lab. Med.*, 3(2), 145 (1995).

163. M. Ventrucci, A. Cipolla, M. Di Stefano, G. M. Ubalducci, M. Middonno, A. Ligabue, and E. Roda, Determination of Fecal Fat Concentration by Near Infrared Spectrometry for the Screening of Pancreatic Steatorrhea, *Int. J. Pancreatol.*, 23(1), 17 (1998).

164. V. Neumeister, J. Henker, G. Kaltenborn, C. Sprössig, and W. Jaross, Simultaneous Determination of Fecal Fat, Nitrogen, and Water by Near-Infrared Reflectance Spectroscopy. *J. Pediatr. Gastroenterol. Nutr.*, 25(4), 388 (1997).

165. T. Nakamura, T. Takeuchi, A. Terada, Y. Tando, and T. Suda, Near-Infrared Spectrometry Analysis of Fat, Neutral Sterols, Bile Acids, and Short-Chain Fatty Acids in the Feces of Patients with Pancreatic Maldigestion and Malabsorption, *Int. J. Pancreatol.*, 23(2), 137 (1998).

166. A. Rivero-Marcotegui, J. E. Olivera-Olmedo, F. S. ValverdeVisus, M. Palacios-Sarrsqueta, A. Grijalba-Uche, and S. Garcia-Merlo, Water, Fat, Nitrogen, and Sugar Content in Feces: Reference Intervals in Children, *Clin. Chem.*, 44(7), 1540 (1998).

167. T. Nakamura, T. Takeuchi, A. Terada, Y. Tando, and T. Suda, Near-Infrared Spectrometry Analysis of Fat, Neutral Sterols, Bile Acids, and Short-Chain Fatty Acids in the Feces of Patients with Pancreatic Maldigestion and Malabsorption, *Int. J. Pancreatol.*, 23(2), 137 (1998).

168. E. Masuko, H. Homma, H. Ohta, S. Nojiri, R. Koyama, and Y. Niitsu, Rheologic Analysis of Gastric Mucosal Hemodynamics in Patients with Cirrhosis, *Gastrointest. Endosc.*, 49(3), 371 (1999).

169. Y. Hirano, K. Omura, H. Yoshiba, N. Ohta, C. Hiranuma, K. Nitta, Y. Nishida, and G. Watanabe, Near-Infrared Spectroscopy for Assessment of Tissue Oxygen Saturation of Transplanted Jejunal Autografts in Cervical Esophageal Reconstruction, *Surg. Today*, 35(1), 67 (2005).

170. J. Kaufman, M. C. Almodovar, J. Zuk, and R. H. Friesen, Correlation of Abdominal Site Near-Infrared Spectroscopy with Gastric Tonometry in Infants Following Surgery for Congenital Heart Disease, Pediatr. *Crit. Care Med.*, 9(1), 62 (2008).

171. T. Noriyuki, H. Ohdan, S. Yoshioka, Y. Miyata, T. Asahara, and K. Dohi, Near-Infrared Spectroscopic Method for Assessing the Tissue Oxygenation State of Living Lung, *Am. J. Respir. Crit. Care Med.*, 156(5), 1656 (1997).

172. G. B. Tirdel, R. Girgis, R. S. Fishman, and J. Theodore, Metabolic Myopathy as a Cause of the Exercise Limitation in Lung Transplant Recipients, *J. Heart Lung Transpl.*, 17(12), 1231 (1998).

173. T. Noriyuki, H. Odan, S. Yoshioka, Y. Miyata, S. Shibata, T. Asahara, Y. Fukuda, and Y. Dohi, Evaluation of Lung Tissue Oxygen Metabolism Using Near-Infrared Spectroscopy, *J. Jpn. Surg. Soc.*, 98(10), 894 (1997).

174. S. Shibata, Y. Miyata, H. Ohdan, T. Noriyuki, S. Yoshioka, S. Okimasa, T. Asahara, H. Ito, and K. Dohi, Estimation of Graft Blood Flow in Rat Lung Transplantation Using Near-Infrared Spectroscopy, *Transplantation Proc.*, 32(7), 2437 (2000).

175. J. S. Tweddell, N. S. Ghanayem, and G. M. Hoffman, Pro: NIRS Is "Standard of Care" for Postoperative Management, *Semin. Thorac. Cardiovasc. Surg. Pediatr. Cardiac Surg. Annu.*, 13(1), 44 (2010).

176. M. Kravari, E. Angelopoulos, I. Vasileiadis, V. Gerovasili, and S. Nanas, Monitoring Tissue Oxygenation during Exercise with Near Infrared Spectroscopy in Diseased Populations—A Brief Review, *Int. J. Ind. Ergon.*, 40(2), 223 (2010).

177. C. D. Kurth, J. L. Steven, L. M. Montenegro, H. M. Watzman, J. W. Gaynor, T. L. Spray, and S. C. Nicolson, Cerebral Oxygen Saturation before Congenital Heart Surgery, *Ann. Thorac. Surg.*, 72(1), 187 (2001).

In vitro blood chemistry

178. M. J. McShane and G. L. Cote, Near-Infrared Spectroscopy for Determination of Glucose, Lactate, and Ammonia in Cell Culture Media, *Appl. Spectrosc.*, 52(8), 1073 (1998).

179. J. Hall, Analysis of Human Breast Milk, *Proc. SPIE*, 3257 (1998).

180. L. Corvaglia, A. Aceti, V. Paoletti, E. Mariani, D. Patrono, G. Ancora, M. G. Capretti, and G. Faldella, Standard Fortification of Preterm Human Milk Fails to Meet Recommended Protein Intake: Bedside Evaluation by Near-Infrared-Reflectance-Analysis, *Early Hum. Dev.*, 86(4), 237 (2010).

181. C. W. Sauer and J. H. Kim, Human Milk Macronutrient Analysis Using Point-of-Care Near-Infrared Spectrophotometry, *J. Perinatol.*, 31(5), 339 (2011).

182. R. A. Shaw, S. Kotowich, H. H. Mantsch, and M. Leroux, Quantitation of Protein, Creatinine, and Urea in Urine by Near-Infrared Spectroscopy, *Clinical Biochem.*, 29(1), 11 (1996).

183. M. Jackson, M. G. Sowa, and H. H. Mantsch, Infrared Spectroscopy: A New Frontier in Medicine, *Biophys. Chem.*, 68, 109 (1997).

184. J. L. Pezzaniti, T.-W. Jeng, L. McDowell, and G. M. Oosta, Preliminary Investigation of Near-Infrared Spectroscopic Measurements of Urea, Creatinine, Glucose, Protein, and Ketone in Urine, *Clin. Biochem.*, 34(3), 239 (2001).

185. S. Tanaka, K. Motoi, M. Nogawa, T. Yamakoshi, and K.-I. Yamakoshi, Feasibility Study of a Urine Glucose Level Monitor for Home Healthcare Using Near Infrared Spectroscopy, in *Annual International Conference of the IEEE Engineering in Medicine and Biology—Proceedings 2006*, p. 6001.

186. W. Liu, W. Yang, L. Liu, and Q. Yu, Use of Artificial Neural Networks in Near-Infrared Spectroscopy Calibrations for Predicting Glucose Concentration in Urine, in *4th International Conference on Intelligent Computing, ICIC 2008*, Shanghai, China, 2008, vol. 5226, p. 1040.

187. S. Tanaka, M. Ogawa, T. Gu, and K.-I. Yamakoshi, Development of Urine Glucose Level Monitor for Home Healthcare Using Near Infrared Spectroscopy, in *8th IEEE International Conference on Bioinformatics and Bioengineering, BIBE 2008*, 2008, article 4696829.

188. K. Kinoshita, H. Morita, M. Miyazaki, N. Hama, H. Kanemitsu, H. Kawakami, P. Wang, O. Ishikawa, H. Kusunoki, and R. Tsenkova, Near Infrared Spectroscopy of Urine Proves Useful for Estimating Ovulation in Giant Panda (*Ailuropoda melanoleuca*), *Anal. Methods*, 2(11), 1671 (2010).

189. K. Kinoshita, M. Miyazaki, H. Morita, M. Vassileva, C. Tang, D. Li, O. Ishikawa, H. Kusunoki, and R. Tsenkova, Spectral Pattern of Urinary Water as a Biomarker of Estrus in the Giant Panda, *Sci. Rep.*, 2 (2012).

190. M. Ogawa, Y. Yamakoshi, K. Motoi, T. Yamakoshi, and K.-I. Yamakoshi, Preliminary Study on Near-Infrared Spectroscopic Measurement of Urine Hippuric Acid for the Screening of Biological Exposure Index, *Proc. SPIE*, 7133 (2009).

191. R. A. Shaw, S. Kotowich, H. H. Eysel, M. Jackson, and G. T. Thomson, Arthritis Diagnosis Based upon the Near-Infrared Spectrum of Synovial Fluid, *Rheumatol. Int.*, 15, 159 (1995).

192. J. M. Zink, J. L. Koenig, and T. R. Williams, Determination of Water Content in Bovine Lenses Using Near-Infrared Spectroscopy, *Ophthalmic Res.*, 29(6), 429 (1997).

193. N. Lopez-Gil, I. Iglasias, and P. Actal, Retinal Image Quality in the Human Eye as a Function of the Accommodation, *Vision Res.*, 38(19), 2897 (1998).

194. R. W. Knighton and X. R. Huang, Visible and Near-Infrared Imaging of the Nerve Fiber Layer of the Isolated Rat Retina, *J. Glaucoma*, 8(1), 317 (1999).

195. D. Lafrance, L. C. Lands, L. Hornby, and D. H. Burns, Near-Infrared Spectroscopic Measurement of Lactate in Human Plasma, *Appl. Spectros.*, 54(2), 300 (2000).

196. J. T. Kuenstner and K. H. Norris, Spectrophotometry of Human Haemoglobin in the Near Infrared Region from 1000 to 2500 nm, *J. Near Infrared Spectrosc.* 2, 595 (1994).

197. J. T. Kuenstner, K. H. Norris, and W. F. McCarthy, Measurement of Hemoglobin in Unlysed Blood by Near-Infrared Spectroscopy, *Appl. Spectrosc.* 48(4), 421 (1994).

198. J. T. Kuenstner and K. H. Norris, Near Infrared Haemoglobinometry, *J. Near Infrared Spectrosc.*, 3, 11 (1995).

199. K. H. Norris and J. T. Kuenstner, Rapid Measurement of Analytes in Whole Blood with NIR Transmittance, in *Leaping Ahead with Near Infrared Spectroscopy*, ed. G. D. Batten, P. C. Flinn, L. A. Welsh, and A. B. Blakeney, NIR Spectroscopy Group, RACI, Australia, 1995, p. 431.

200. D. Lafrance, L. C. Lands, and D. H. Burns, In Vivo Lactate Measurement in Human Tissue by Near-Infrared Diffuse Reflectance Spectroscopy, *Vib. Spectrosc.*, 36(2), 195 (2004).

201. H. M. Schipper, C. S. Kwok, S. M. Rosendahl, D. Bandilla, O. Maes, C. Melmed, D. Rabinovitch, and D. H. Burns, Spectroscopy of Human Plasma for Diagnosis of Idiopathic Parkinson's Disease, *Biomarkers Med.*, 2(3), 229 (2008).

202. D. H. Burns, S. Rosendahl, D. Bandilla, O. C. Maes, H. M. Chertkow, and H. M. Schipper, Near-Infrared Spectroscopy of Blood Plasma for Diagnosis of Sporadic Alzheimer's Disease, *J. Alzheimers Dis.*, 17(2), 391 (2009).

203. S. Zhang, In Vivo Determination of Myocardial pH during Regional Ischemia Using NIR Spectroscopy, *Proc. SPIE*, 3257 (1998).

204. D. Burns, NIR Methods for Health Assessment: New Eyes on an Old Problem, presented at EAS 2013, Somerset, NJ.

205. D. Burns, Near Infrared Spectroscopy for Assessment of Fetal, Maternal and Infant Health, presented at EAS 2013, Somerset, NJ.

Newborns

206. K. D. Liem and G. Greisen, Monitoring of Cerebral Haemodynamics in Newborn Infants (Review), *Early Hum. Dev.*, 86(3), 155 (2010).
207. B. Uriesberger, A. Kaspirek, G. Pichler, and W. Muller, Apnoea of Prematurity and Changes in Cerebral Oxygenation and Cerebral Blood Volume, *Neuropediatrics*, 30(1), 29 (1999).
208. W. S. Park and Y. S. Chang, Effects of Decreased Cerebral Perfusion Pressure on Cerebral Hemodynamics, Brain Cell Membrane Function, and Energy Metabolism during the Early Phase of Experimental *Escherichia coli* Meningitis in the Newborn Piglet, *Neurological Res.*, 21(3), 345 (1999).
209. Y. S. Chang, W. S. Park, M. Lee, K. S. Kim, S. M. Shin, and J. H. Choi, Near-Infrared Spectroscopic Monitoring of Secondary Cerebral Energy Failure after Transient Global Hypoxia-Ischemia in the Newborn Piglet, *Neurological Res.*, 21(2), 216 (1999).
210. V. Dietz, M. Wolf, M. Keel, K. V. Siebenthal, O. Baenziger, and H. Bucher, CO_2 Reactivity of the Cerebral Hemoglobin Concentration in Healthy Term Newborns Measured by Near-Infrared Spectrometry, *Biol. Neonate*, 75(2), 85 (1999).
211. F. F. Buchvald, K. Kesje, and G. Greisen, Measurement of Cerebral Oxyhaemoglobin Saturation and Jugular Blood Flow in Term Healthy Newborn Infants by Near-Infrared Spectroscopy and Jugular Venous Occlusion, *Biol. Neonate*, 75(2), 97 (1999).
212. C. Roll, J. Knief, S. Horsch, and L. Hanssler, Effect of Surfactant Administration on Cerebral Haemodynamics and Oxygenation in Premature Infants—A Near-Infrared Spectroscopy Study, *Neuropediatrics*, 31(1), 16 (2000).
213. C. Dani, G. Bertini, M. F. Reali, M. Tronchin, L. Wiechmann, E. Martelli, and F. F. Rubaltelli, Brain Hemodynamic Changes in Preterm Infants after Maintenance Dose Caffeine and Aminophylline Treatment, *Biol. Neonate*, 78(1), 27 (2000).
214. A. J. Petros, R. Hayes, R. C. Tasker, P. M. Fortune, I. Roberts, and E. Kiely, Near-Infrared Spectroscopy Can Detect Changes in Splanchnic Oxygen Delivery in Neonates during Apnoeic Episodes, *Eur. J. Pediatr.*, 158(2), 173 (1999).
215. I. A. Hassan, S. A. Spencer, Y. A. Wickramasinghe, and K. S. Palmer, Measurement of Peripheral Oxygen Utilisation in Neonates Using Near Infrared Spectroscopy: Comparison between Arterial and Venous Occlusion Methods, *Early Hum. Dev.*, 57(3), 211 (2000).
216. M. Wolf, G. Naulaers, F. V. Bel, S. Kleiser, and G. Greisen, A Review of Near Infrared Spectroscopy for Term and Preterm Newborns, *J. Near Infrared Spectros.*, 20(1), 43 (2012).
217. P. Giliberti, V. Mondì, A. Conforti, M. Haywood Lombardi, S. Sgrò, P. Bozza, S. Picardo, and P. Bagolan, Near Infrared Spectroscopy in Newborns with Surgical Disease, *J. Maternal-Fetal Neonatal Med.*, 24(Suppl 1), 56 (2011).
218. B. A. Johnson, G. M. Hoffman, J. S. Tweddell, J. R. Cava, M. Basir, M. E. Mitchell, M. C. Scanlon, K. A. Mussatto, and N. S. Ghanayem, Near-Infrared Spectroscopy in Neonates before Palliation of Hypoplastic Left Heart Syndrome, *Ann. Thorac. Surg.*, 87, 571 (2009).

219. C. A. Reed, R. S. Baker, C. T. Lam, J. L. Hilshorst, R. Ferguson, J. Lombardi, and P. Eghtesady, Application of Near-Infrared Spectroscopy during Fetal Cardiac Surgery, *J. Surg. Res.*, 171, 159 (2011).

220. M. Ranger, C. C. Johnston, C. Limperopoulos, and J. Rennick, Cerebral Near-Infrared Spectroscopy as a Measure of Nociceptive Evoked Activity in Critically Ill Infants, *Pain Res. Manag. J. Can. Pain Soc.*, 16, 5 (2011).

221. R. G. Wijbenga, P. M. A. Lemmers, and F. Van Bel, Cerebral Oxygenation during the First Days of Life in Preterm and Term Neonates: Differences between Different Brain Regions, *Pediatr. Res.*, 70, 389 (2011).

222. S. Aoyama, T. Toshima, Y. Saito, N. Konishi, K. Motoshige, N. Ishikawa, K. Nakamura, and M. Kobayashi, Maternal Breast Milk Odour Induces Frontal Lobe Activation in Neonates: A NIRS Study, *Early Hum. Dev.*, 86(9), 541 (2010).

223. D. D. Infante, O. O. Segarra, S. S. Redecillas, M. M. Alvarez, and M. M. Miserachs, Modification of Stool's Water Content in Constipated Infants: Management with an Adapted Infant Formula, *Nutr. J.*, 10(1) (2011).

224. K.-Z. Liu, R. A. Shaw, T. C. Dembinski, G. J. Reid, S. L. Ying, and H. H. Mantsch, Comparison of Infrared Spectroscopic and Fluorescence Depolarization Assays for Fetal Lung Maturity, *Am. J. Obstet. Gynecol.*, 183(1), 181 (2000).

225. K. M. Power, J. E. Sanchez-Galan, G. W. Luskey, K. G. Koski, and D. H. Burns, Use of Near-Infrared Spectroscopic Analysis of Second Trimester Amniotic Fluid to Assess Preterm Births, *J. Pregnancy*, 2011, 980 (2011).

226. G. Vishnoi, A. H. Hielscher, N. Ramanujam, and B. Chance, Photon Migration through Fetal Head In Utero Using Continuous Wave, Near-Infrared Spectroscopy: Development and Evaluation of Experimental and Numerical Models, *J. Biomed. Opt.*, 5(2), 163 (2000).

227. N. Ramanujam, H. Long, M. Rode, I. Forouzan, M. Morgan, and B. Chance, Antepartum, Transabdominal Near-Infrared Spectroscopy: Feasibility of Measuring Photon Migration through the Fetal Head In Utero, *J. Maternal-Fetal Med.*, 8(6), 275 (1999).

228. C. J. Calvano, M. E. Moran, B. A. Mehlhaff, B. L. Sachs, and J. Mandell, Amnioscopic Endofetal Illumination with Infrared-Guided Fiber, *J. Endourol.*, 11(4), 259 (1997).

Cancer and precancer

229. Z. Ge et al., Screening PAP Smears with Near-Infrared Spectroscopy, *Appl. Spectros.*, 49(4), 1324 (1995).

230. Y. Yang, Q. Hou, H. Zhao, Z. Qin, and F. Gao, The Cervical Cancer Detection System Based on an Endoscopic Rotary Probe, *Proc. SPIE*, 8214 (2012).

231. Y. Xiang, J. Tian, Z. Zhang, and Y. Dai, Diagnosis of Endometrial Cancer Based on Near Infrared Spectroscopy and General Regression Neural Network, in *Proceedings—2010 6th International Conference on Natural Computation, ICNC 2010*, 2010, vol. 3, p. 1228.

232. V. R. Kondepati, M. Keese, R. Mueller, B. C. Manegold, and J. Backhaus, Application of Near-Infrared Spectroscopy for the Diagnosis of Colorectal Cancer in Resected Human Tissue Specimens, *Vib. Spectrosc.*, 44, 236 (2007).

233. V. R. Kondepati, M. Keese, R. Mueller, and J. Backhaus, Near-Infrared Spectroscopic Detection of Human Colon Diverticulitis: A Pilot Study, *Vib. Spectrosc.*, 44, 56 (2007).

234. V. R. Kondepati, T. Oszinda, H. M. Heise, K. Luig, R. Mueller, O. Schroeder, M. Keese, and J. Backhaus, CH-Overtone Regions as Diagnostic Markers for Near-Infrared Spectroscopic Diagnosis of Primary Cancers in Human Pancreas and Colorectal Tissue, *Anal. Bioanal. Chem.*, 387, 1633 (2007).

235. X. Shao, W. Zheng, and Z. Huang, Near-Infrared Autofluorescence Spectroscopy for In Vivo Identification of Hyperplastic and Adenomatous Polyps in the Colon, *Biosensors Bioelectronics*, 30, 118 (2011).

236. V. Ntziachristos et al., *Simultaneous MRI and NIR Mammographic Examination*, University of Pennsylvania, Philadelphia, www.1rsm.upenn.edu/~vasilis/Concurrent.html.

237. Optical Biopsy Would Be Fast, Painless, and Inexpensive, *Science Daily*, July 13, 1999, www.sciencedaily.com.

238. B. W. Pogue, S. P. Poplack, T. O. McBride, W. A. Wells, K. S. Osterman, U. L. Osterberg, and K. D. Paulsen, Quantitative Hemoglobin Tomography with Diffuse Near-Infrared Spectroscopy: Pilot Results in the Breast, *Radiology*, 218, 261 (2001).

239. Y. Gu, R. Mason, and H. Liu, Estimated Fraction of Tumor Vascular Blood Contents Sampled by Near Infrared Spectroscopy and 19F Magnetic Resonance Spectroscopy, *Opt. Express*, 13(5), 1726.

240. Y. Song, J. G. Kim, R. P. Mason, and H. Liu, Investigation of Rat Breast Tumour Oxygen Consumption by Near-Infrared Spectroscopy, *J. Phys. D Appl. Phys.*, 38, 2682 (2005).

241. H. B. Stone, J. M. Brown, T. L. Phillips, and R. M. Sutherland, Oxygen in Human Tumors: Correlations between Methods of Measurement and Response to Therapy, *Radiat. Res.*, 136(3), 422 (1993).

242. Y. Gu, W. R. Chen, M. Xia, S. W. Jeong, and H. Liu, Effect of Photothermal Therapy on Breast Tumor Vascular Content: Noninvasive Monitoring by Near-Infrared Spectroscopy, *Photochem. Photobiol.*, 81, 4 (2005).

243. V. Ntziachristos et al., *Simultaneous MR and NIR Mammographic Examination*, University of Pennsylvania, Philadelphia, 1997, www.1rsm.upenn.edu/~vasilis/frresearch.html.

244. V. Saxena, I. Gonzalez-Gomez, and W. E. Laug, A Noninvasive Multimodal Technique to Monitor Brain Tumor Vascularization, *Phys. Med. Biol.*, 52, 5295 (2007).

245. C. M. Carpenter, S. Srinivasan, B. W. Pogue, and K. D. Paulsen, Methodology Development for Three-Dimensional MR-Guided Near Infrared Spectroscopy of Breast Tumors, *Opt. Express*, 16(22), 1790 (2008).

246. P. J. Milne, J. M. Parel, D. B. Denham, X. Gonzalez-Cirre, and D. S. Robinson, Development of a Stereotactically Guided Laser Interstitial Thermotherapy of Breast Cancer In Situ Measurement and Analysis of the Temperature Field in Ex Vivo and In Vivo Adipose Tissue, *Lasers Surg. Med.*, 26(1), 67 (2000).

247. B. J. Tromberg, N. Shah, R. Lanning, A. Cerussi, J. Espinoza, T. Pham, L. Svaasand, and J. Butler, Non-Invasive In Vivo Characterization of Breast Tumors Using Photon Migration Spectroscopy, *Neoplasia*, 2(1–2), 26 (2000).

248. T. D. Tosteson, B. W. Pogue, E. Demidenko, T. O. McBride, and K. D. Paulsen, Confidence Maps and Confidence Intervals for Near-Infrared Images in Breast Cancer, *IEEE Trans. Med. Imaging*, 18(12), 1188 (1999).
249. E. L. Hull, D. L. Conover, and T. H. Foster, Carbogen Induced Changes in Rat Mammary Tumour Oxygenation Reported by Near-Infrared Spectroscopy, *Br. J. Cancer*, 79(11–12), 1709 (1999).
250. M. Meurens et al., Identification of Breast Carcinomatous Tissue by Near-Infrared Reflectance Spectroscopy, *Appl. Spectrosc.*, 48(2), 190 (1994).
251. T. Jarm, Y. A. Wickramasinghe, M. Deakin, M. Cemazar, J. Elder, P. Rolfe, G. Sersa, and D. Miklavcic, Blood Perfusion of Subcutaneous Tumours in Mice Following the Application of Low-Level Direct Electric Current, *Adv. Exp. Med. Biol.*, 471, 497 (1999).
252. J. Krohn, P. Svenmarker, C. T. Xu, S. J. Mørk, and S. Andersson-Engels, Transscleral Optical Spectroscopy of Uveal Melanoma in Enucleated Human Eyes, *Invest. Ophthalmol. Vis. Sci.*, 53(9), 5379 (2012).
253. U. Mahmood, C. H. Tung, A. Bogdanov, and R. Weissleder, Near Infrared Optical Imaging of Protease Activity for Tumor Detection, *Radiology*, 213, 833 (1999).
254. C.-H. Tung, Y. Lin, W. K. Moon, and R. Weissleder, A Receptor-Targeted Near-Infrared Fluorescence Probe for In Vivo Tumor Imaging *ChemBioChem*, 8, 784 (2002).
255. I. Hilger Y. Leistner, A. Berndt, C. Fritsche, K. Michael, H. Hartwig, K. Werner, and A. Kaiser, Near-Infrared Fluorescence Imaging of HER-2 Protein Over-Expression in Tumour Cells, *Eur. Radiol.*, 14, 1124 (2004).
256. Y. Chen, X. Intes, and B. Chance, Development of High-Sensitivity Near-Infrared Fluorescence Imaging Device for Early Cancer Detection, *Biomed. Instrum. Technol.*, 39(1), 76 (2005).
257. Y. Wu, W. Cai, and X. Chen, Near-Infrared Fluorescence Imaging of Tumor Integrin $\alpha v \beta 3$ Expression with Cy7-Labeled RGD Multimers, *Mol. Imaging Biol.*, 8, 226 (2006).
258. M. Meincke, S. Tiwari, K. Hattermann, H. Kalthoff, and R. Mentlein, Near-Infrared Molecular Imaging of Tumors via Chemokine Receptors CXCR4 and CXCR7, Photon Diffusion, Absorption, and Scattering, *Clin. Exp. Metastasis*, 28, 713 (2011).

Photon migration in tissues

259. S. L. Jacques, Optical Properties of Biological Tissues: A Review, *Phys. Med. Biol.*, 58, R37(2013).
260. R. R. Anderson and J. A. Parrish, Optical Properties of Human Skin, in *The Science of Photomedicine*, ed. J. D. Regan and J. A. Parrish, Plenum Press, New York, 1982, p. 147.
261. P. Parsa, S. L. Jacques, and N. Nishioka, Optical Properties of the Liver between 350 and 2200 nm, *Appl. Opt.* 28, 2325 (1989).
262. V. G. Peters, D. R. Wymant, M. S. Patterson, and G. L. Frank, Optical Properties of Normal and Diseased Human Breast Tissues in the Visible and Near Infrared, *Phys. Med. Biol.*, 35, 1317 (1990).

263. H. Firbank, M. Hiraoka, M. Essenpreis, and D. T. Delpy, Measurement of the Optical Properties of the Skull in the Wavelength Range 650–950 nm, *Phys. Med. Biol.*, 38, 503 (1993).

264. S. R. Arridge, M. Schweiger, M. Hiraoka, and D. T. Delpy, A Finite Element Approach for Modeling Photon Transport in Tissue, *Med. Phys.*, 20(2), 299 (1993).

265. B. Lin, V. Chernomordik, A. Gandjbakhche, D. Matthews, and S. Demos, Investigation of Signal Dependence on Tissue Thickness in Near Infrared Spectral Imaging, *Opt. Express*, 15(25), 16581 (2007).

266. P. Sadoghi, Influence of Scattering, Tissue Optical Parameters and Interface Reflectivities on Photon Migration in Human Tissue, *J. Modern Opt.*, 54(6), 845 (2007).

267. E. Alerstam, S. Andersson-Engels, and T. Svensson, White Monte Carlo for Time-Resolved Photon Migration, *J. Biomed. Opt.*, 13(4) (2008).

268. F. Martelli, S. Del Bianco, G. Zaccanti, A. Liemert, A. Kienle, M. Schweiger, S. R. Arridge, S. Prerapa, A. Jelzow, H. Wabnitz, N. Zolek, and A. Liebert, Progress in Biomedical Optics and Imaging, *Proc. SPIE*, 7896 (2011).

269. J. Kikuta and M. Ishii, Recent Advances in Intravital Imaging of Dynamic Biological Systems, *J. Pharmacol. Sci.*, 119(3), 193 (2012).

270. Y. P. Chen, X. Ma, C. B. Li, and S. Q. Chen, Research on the Near-Infrared (NIR) Photon Migration in the Knee, *Adv. Mater. Res.*, 760–762, 388 (2013).

271. V. P. Hart and T. E. Doyle, Simulation of Diffuse Photon Migration in Tissue by a Monte Carlo Method Derived from the Optical Scattering of Spheroids, *Appl. Opt.*, 52(25), 6220 (2013).

272. C. E. W. Gributs and D. H. Burns, Multiresolution Analysis for Quantification of Optical Properties in Scattering Media Using Pulsed Photon Time-of-Flight Measurements, *Anal. Chim. Acta*, 490(1–2), 185 (2003).

273. C. E. W. Gributs and D. H. Burns, Fractal Dimension Analysis of Time-Resolved Diffusely Scattered Light from Turbid Samples, *Anal. Chem.*, 77(13), 4213 (2005).

Review articles

274. S. K. Vashist, Non-Invasive Glucose Monitoring Technology in Diabetes Management: A Review, *Anal. Chim. Acta*, 750, 16 (2012).

275. C.-F. So, K.-S. Choi, T. K. S. Wong, and J. W. Y. Chung, Recent Advances in Noninvasive Glucose Monitoring, *Med. Dev. Evid. Res.*, 5(1), 45 (2012).

276. M. Ferrari and V. Quaresima, Review: Near Infrared Brain and Muscle Oximetry: From the Discovery to Current Applications, *J. Near Infrared Spectros.*, 1(14), 20 (2012).

277. B. R. Soller, J. Sliwa, Y. Yang, F. Zou, K. L. Ryan, C. A. Rickards, and V. A. Convertino, Simultaneous Spectroscopic Determination of Forearm Muscle pH and Oxygen Saturation during Simulated Haemorrhage, *J. Near Infrared Spectros.*, 20, 141 (2012).

278. D. A. Hampton and M. A. Schreiber, Near Infrared Spectroscopy: Clinical and Research Uses, *Transfusion*, 53(Suppl 1), 52S (2013).

279. Y. Bhambhani, Review: Application of Near Infrared Spectroscopy in Evaluating Cerebral and Muscle Haemodynamics during Exercise and Sport, *J. Near Infrared Spectros.*, 20, 117 (2012).

280. D. Contini, L. Zucchelli, L. Spinelli, M. Caffini, R. Re, A. Pifferi, R. Cubeddu, and A. Torricelli, Review: Brain and Muscle Near Infrared Spectroscopy/ Imaging Techniques *J. Near Infrared Spectros.*, 20, 15 (2012).
281. L. C. Enfield and A. P. Gibson, Review: A Review of Mechanisms of Contrast for Diffuse Optical Imaging of Cancer, *J. Near Infrared Spectros.*, 20, 185 (2012).
282. A. Pifferi, A. Farina, A. Torricelli, G. Quarto, R. Cubeddu, and P. Taronia, Review: Time-Domain Broadband Near Infrared Spectroscopy of the Female Breast: A Focused Review from Basic Principles to Future Perspectives, *J. Near Infrared Spectros.*, 20, 223 (2012).
283. D. Grosenick, H. Wabnitz, and B. Ebert, Review: Recent Advances in Contrast-Enhanced Near Infrared Diffuse Optical Imaging of Diseases Using Indocyanine Green, *J. Near Infrared Spectros.*, 20, 203 (2012).
284. F. Martelli, Review: An ABC of Near Infrared Photon Migration in Tissues: The Diffusive Regime of Propagation, *J. Near Infrared Spectros.*, 20, 29 (2012).

Index